D0143353

A programmer's geometry

Adrian Bowyer
BSc, PhD, ACGI, MBCS
University of Bath

and

John Woodwark
BSc, PhD, CEng, MIMechE, MBCS
University of Bath

Butterworths
London Boston Durban Singapore Sydney Toronto Wellington

All rights reserved. No part of this publication may be reproduced or transmitted in any form or by any means, including photocopying and recording, without the written permission of the copyright holder, applications for which should be addressed to the Publishers. Such written permission must also be obtained before any part of this publication is stored in a retrieval system of any nature.

This book is sold subject to the Standard Conditions of Sale of Net Books and may not be re-sold in the UK below the net price given by the Publishers in their current price list.

First published, 1983
Reprinted, 1984, 1985

© A Bowyer & J R Woodwark, 1983

British Library Cataloguing in Publication Data

Bowyer, A.
A programmer's geometry.
1. Geometry
I. Title   II. Woodwark, J.
516   QA445

ISBN 0-408-01242-0 Pbk

Printed in England by Butler and Tanner Ltd., Frome and London

A straight line segment can be drawn joining any two points.

Any straight line segment can be extended indefinitely in a straight line.

Given any straight line segment, a circle may be drawn having the segment as radius and one end point as centre.

All right angles are congruent.

If two lines are drawn which intersect a third in such a way that the sum of the inner angles on one side is less than two right angles, then the two lines inevitably intersect on that side if they are extended far enough.

Euclid's five postulates
(circa 300 B.C.)

To describe right lines and circles are problems, but not geometrical problems. The solution of these problems is required from mechanics, and by geometry the use of them, when so solved, is shown; and it is the glory of geometry that, from those few principles brought from without, it is able to produce so many things.

Isaac Newton's preface to Principia
(1686 A.D.)

Foreword

This book aims to fulfil two needs of the programmer whose work includes geometric calculations. Firstly, it provides useful formulae in a single source. Secondly, it presents advice on, and examples of, the computer representation and coding of geometry. Applications are not restricted to computer graphics, although graphics are useful for checking, even when the working program will produce no pictures.

The reader is assumed to have a grasp of simple Euclidean geometry and the Cartesian coordinate system. For some sections an elementary knowledge of calculus and vector algebra is desirable. The reader is also assumed to be familiar with a scientific programming language, such as FORTRAN, Pascal, or BASIC.

The scope of this book has been determined by the authors' own experiences in designing and implementing programs. We hope that many readers will find that it is applicable to their work, too. We would, however, be interested to hear about material that we should, perhaps, have included. Notification of mistakes that have escaped checking will also be received with thanks and acknowledgement.

A number of the authors' colleagues have assisted in the preparation of this book. We would like to thank them all, and in particular Professor John Fitch and Julian Padget for the use of their symbolic algebra package and Dr Peter Green as the originator of the subroutines for perspective projection described in Chapter Six. We are also grateful for the understanding shown by our wives.

Adrian Bowyer and John Woodwark

University of Bath, Autumn 1982

Contents

Introduction 1

1 Points and lines 5

1.1 Distance between Two Points 5
1.2 Equations of a Line 7
1.3 Distance from a Point to a Line 12
1.4 Angle between Two Lines 13
1.5 Intersection of Two Lines 16
1.6 Line through Two Points 19
1.7 Line Equidistant from Two Points 20
1.8 Normal to a Line through a Point 21

2 Points, lines and circles 23

2.1 Equations of a Circle 23
2.2 Intersections of a Line and a Circle 25
2.3 Intersections of Two Circles 27
2.4 Tangents from a Point to a Circle 29
2.5 Tangents to a Circle Normal to a Line 31
2.6 Tangents between Two Circles 33
2.7 Circles of Given Radius through Two Points 35
2.8 Circles of Given Radius through a Point and Tangent to a Line 36
2.9 Circles of Given Radius Tangent to Two Lines 38
2.10 Circles of Given Radius through a Point and Tangent to a Circle 40
2.11 Circles of Given Radius Tangent to a Line and a Circle 42
2.12 Circles of Given Radius Tangent to Two Circles 45

3 Points, line segments and arcs 46

3.1 Representation of a Line Segment 46
3.2 Distance from a Point to a Line Segment 47
3.3 Intersection of Two Line Segments 48
3.4 Representation of an Arc 53
3.5 Distance from a Point to an Arc 56
3.6 Intersections of a Line Segment and an Arc 58
3.7 Intersections of Two Arcs 61

4 Areas 62

4.1 Area of a Triangle 62
4.2 Centre of Gravity of a Triangle 63
4.3 Incentre of a Triangle 64
4.4 Circumcentre of a Triangle 65
4.5 Representation of a Polygon 66
4.6 Area of a Polygon 69
4.7 Centre of Gravity of a Polygon 70
4.8 Centre of Gravity of a Sector and a Segment 72

5 Curves other than circles 74

5.1 General Implicit Quadratic Equations 74
5.2 Interpolation Using General Implicit Quadratics 75
5.3 Parametric Polynomials 76
5.4 Interpolation Using Parametric Polynomials 77
5.5 Parametric Spline Curves 79
5.6 Radius of Curvature 82

6 Vectors, matrices and transformations 83

6.1 Vectors 83
6.2 Matrices 86
6.3 Determinants 87
6.4 Transformations 88
6.5 Perspective 90

7 Points, lines and planes 96

7.1 Distance between Two Points in Space 96
7.2 Equations of a Straight Line in Space 97
7.3 Distance from a Point to a Line in Space 98
7.4 Distance between Two Lines in Space 100
7.5 Angle between Two Lines in Space 101
7.6 Line through Two Points in Space 103
7.7 Equation of a Plane 104
7.8 Distance from a Point to a Plane 107
7.9 Angle between a Line and a Plane 108
7.10 Angle between two Planes 110
7.11 Intersection of a Line and a Plane 111
7.12 Intersection of Three Planes 112
7.13 Intersection of Two Planes 113
7.14 Plane through Three Points 114
7.15 Plane through a Point and Normal to a Line 116
7.16 Plane through Two Points and Parallel to a Line 116

8 Volumes 118

8.1 Volume of a Tetrahedron 118
8.2 Centre of Gravity and Surface Area of a Tetrahedron 119
8.3 Circumcentre of a Tetrahedron 120
8.4 Volume and Surface Area of a Cylinder and of a Sphere 122
8.5 Volume and Centre of Gravity of a Sector of a Sphere 123
8.6 Volume, Surface Area and Centre of Gravity of a Cone 124

9 Drawing pictures 126

9.1 Line and Pixel Devices 126
9.2 Circles on Line Devices 127
9.3 Arcs on Line Devices 129
9.4 Lines on Pixel Devices 130
9.5 Circles on Pixel Devices 131
9.6 Clipping 133

References 138

Introduction

The chapters in this book are divided into sections, each dealing with the representation and solution of a single geometrical problem. The sections are usually headed with a diagram to make reference easier. As far as possible uniform conventions have been observed throughout diagrams, algebra, and code. The nomenclature adopted is a compromise between uniformity and the symbols already in wide use for certain equations. It is as follows:

A, B, C, and D	Coefficients of implicit equations of lines and planes
F, G, and H	Coefficients of parametric lines
I	Subscript of a known radius
J, K, L, M, and N	Labels of points
P and Q	Vectors
R	Radius and distance in general
S and T	Parameters
U, V, and W	Second set of coordinate values
X, Y, and Z	Coordinates in space
α, β, γ, and θ	Angles (in radians)

In the text and diagrams only points appear as capital letters. In the coding examples capitals are used throughout to avoid problems for readers whose computer facilities do not support lower case letters. Symbols are never used for different purposes in upper and lower case.

Subscripts are widely employed in a number of different ways. They may define more than one set of constants, as in two different straight lines:

$$a_1x + b_1y + c_1 = 0$$

$$a_2x + b_2y + c_2 = 0$$

They may specify values corresponding to points, so that, for example, the coordinates of point J would be (x_J, y_J), or they may indicate parameter values; x_0 might be the value of x where a parameter, t, say, was zero. Generally, subscripted variables are those known in advance, and variables without subscripts are to be calculated. Primed variables, such as x', are avoided as far as possible, but are sometimes used where numerical subscripts would be inappropriate.

In drawing the diagrams the aim has been to strike a balance between consistency and an excessive use of special symbols, line types, and tones. Three line types (dashed, thin, and thick) show lines of increasing interest, and dashed lines are also used where only distance, rather than an actual line, is to be indicated. Three intensities of tone are used to define areas, and to indicate surfaces and solids. Points given as part of a problem are labelled with the letters J to N, as mentioned above, while unknown points are labelled (x, y), with numerical subscripts on x and y when there is more than one value for each. Lines are unlabelled if they are given, unless there is more than one, when they are numbered. Lines to be calculated are labelled with their equation: $ax + by + c = 0$. Planes are labelled similarly. After the first diagram in a section, which usually shows a simple case of the geometry to be discussed, the conventions are often relaxed, especially in complicated diagrams enumerating all possible cases of a problem. This informality in labelling also extends to non-circular curves and three-dimensional figures, with clarity as the main objective throughout.

The selection of a programming language in which to provide examples of coding involved some deliberation. The authors finally decided to use FORTRAN 77. They did this in an attempt to recognise the obvious virtues of structured languages, whilst acknowledging that many geometric applications are tied to a vast amount of existing software, such as graphics packages, written in FORTRAN. They hope that there is enough structure in FORTRAN 77 to satisfy ALGOL 68 and C programmers. In any case there are no curly brackets or semi-colons to baffle the BASIC enthusiast. The authors have tried to use FORTRAN 77 straightforwardly, too. All the algebra has been written out in the code, without attempting to use arrays, subroutines, or functions to mimic the vector and matrix operations available implicitly in other languages, such as APL.

The code is distinguished from the rest of the text and algebra by a different typeface. Comments have not been included in FORTRAN standard form, but in italics to the right of lines of code, to save space. The number of comments has been decided by the fact that code is preceded by a diagram and explanation. The reader would often be well advised to lay his code out more sparsely with more comments, directed, of course, at his particular problem. Most of the examples of code are assumed to be part of a larger program section, not subroutines in their own right. In these cases there are no SUBROUTINE, FUNCTION, RETURN, or END statements. Any arrays used are declared in a disconnected section above the executable statements.

If a condition, most often an error, occurs such that all the code should not be executed, a special italic comment line is inserted, starting with a row of dots. This indicates what condition has occurred, and should be replaced in an actual program by code to report the error in a WRITE statement or to set a condition flag. As a final deviation from the FORTRAN 77 standard, continuation lines are started with the ampersand (&) character, which is a little more readable than the FORTRAN standard system when column numbers are not shown.

Single variables from the text are translated directly into the code. Subscripts simply form the second letter of the variable name, so that x_J becomes XJ, for example. Angles are written out as ALPHA, BETA, GAMMA and THETA, and are never subscripted. When additional variables are introduced, these are assembled as far as possible from the components of the relevant algebraic expression. Thus $x_L - x_K$ becomes XLK, for instance. When two different terms would have the same name using this convention, or the FORTRAN six letter name limit is exhausted, other mnemonics are chosen. Some names, such as DENOM for the bottom line of a fraction, and ROOT for the result of a square root operation, are used consistently throughout. Also suffices, such as SQ for a squared term, or INV for a reciprocal, are generally employed, again where the name length restriction will allow. Thus lines such as

```
XJSQ = XJ*XJ
XJINV = 1.0/XJ
```

are common. Multiplication by a reciprocal is often used when many terms must be divided by a single term. This is done because the division operation is commonly the slowest on a computer, but the best balance will depend on the reader's own system.

If there is any chance at all of the denominator in a division operation being so near to zero that the result of the division exceeds the largest real number with which the computer can cope, then the value of the denominator must be checked before the division commences. In this book this is achieved by comparing the absolute value of the denominator with a notional accuracy value, which is always called ACCY. This value should be selected with regard for the likely size of numbers that will be encountered in a program. ACCY should be set so as to avoid rejecting good data, while also considering the characteristics of the computer being used. For instance, an accuracy parameter of 10^{-6} might be set in a data statement:

```
DATA ACCY /1.0E-6/
```

This is a good general purpose value for systems where real number calculations are done in a floating point format giving an average resolution of 6 or 7 decimal places, and the data are values not too far from 1 (say 0.01 to 100.0). In extreme cases a single accuracy parameter may not be applicable throughout a program, especially if it is used for other purposes; for instance to determine the distance between points below which they can be considered to be coincident. Careful thought is needed to decide what accuracy values actually refer to, particularly when values to be compared are dimensionally different, such as distance and squared distance. These are problems of numerical analysis, and the reader is referred to Weeg and Reed for a fuller treatment.

Where other books are referred to in the text only the author's name is given. There is a list of references after Chapter Nine, where the reader may find the full titles of the books, together with a few words about each one.

1
Points and lines

1.1 Distance between Two Points

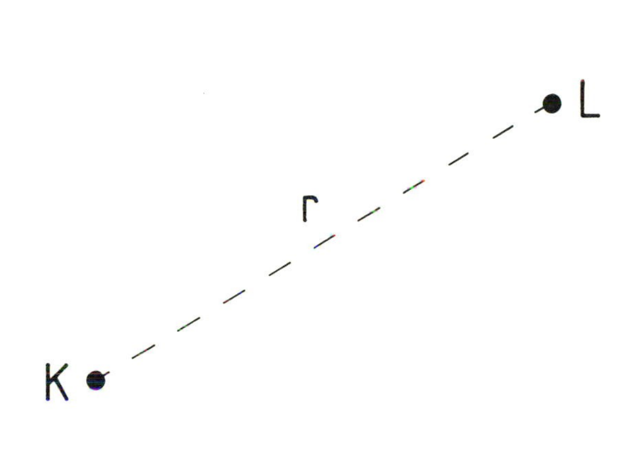

Given two points, K and L, an elementary application of Pythagoras' theorem gives the distance between them as

$$r = \sqrt{[(x_L - x_K)^2 + (y_L - y_K)^2]}$$

which is simply coded:

```
XLK = XL - XK
YLK = YL - YK
RSQ = XLK*XLK + YLK*YLK
R = SQRT(RSQ)
```

However, this is not a cheap expression to compute, and many applications require a very large number of distance calculations. In most cases, these are repeated *comparisons* of inter-point distances with a reference distance. This reference distance is commonly either a fixed value, or a minimum or maximum that is being updated as an algorithm proceeds.

The simplest way of increasing the efficiency of comparisons is to compare values of r^2 instead of r. A reference distance must, of course, be squared before comparisons start, and it may be advantageous to use squared distances throughout a data structure.

If a large number of comparisons are being made, of which many will lead to the rejection of distances grossly outside the distance of interest, then pre-testing can be performed which will avoid the computation of squares as well as the square root for these cases.

If $|x_K - x_L|$ or $|y_K - y_L|$ is greater than a reference distance, then r must be as well, and further comparison can be avoided:

```
XLK = XL - XK
IF (ABS(XLK).GE.RREF) THEN

        ..... don't bother to continue.

ELSE
        YLK = YL - YK
        IF (ABS(YLK).GE.RREF) THEN

                ..... don't bother to continue.

        ELSE
                RSQ = XLK*XLK + YLK*YLK
                IF (RSQ.GE.RREFSQ) THEN

                        ..... don't bother to continue.

                ELSE

                        ..... RSQ is within the square of the
                           reference distance.

                ENDIF
        ENDIF
ENDIF
```

Note that both the reference distance, r_{ref}, and its square must be stored (and possibly, depending on the problem, updated).

If, on the other hand, the requirement is to reject quickly distances less than the reference distance, then we can use the fact that r must be less than $|x_L - x_K| + |y_L - y_K|$.

If the points under consideration are known to be (or are suspected of being) on a grid, then it may be worth dealing with the trivial cases where the two points are on the same horizontal or vertical grid line separately.

Finally, when working out the square root of the sum of two squares, it is slightly less efficient, but more numerically stable to use the code

```
TEMP1 = ABS(XLK)                   Temporary store for
TEMP2 = ABS(YLK)                   absolute values.

AMN = AMIN1(TEMP1,TEMP2)           Find the smaller and
AMX = AMAX1(TEMP1,TEMP2)           the larger.

DIV = AMN/AMX                      Prevent terms in the
R = AMX*SQRT(1.0+DIV*DIV)          bracket getting too big.
```

than the code given near the beginning of this section. This is probably only worth bothering with if you have very large and very small distances to deal with. This point is worth remembering throughout the rest of the book, however, as square rooting a sum of squares often needs to be done, and, though we use the simpler form, the above code might be needed in exceptional cases.

1.2 Equations of a Line

The choice of algebraic expressions to represent straight lines is important. Many operations are algebraically considerably simpler with one form than with another, and hence yield shorter and quicker code. In general one form should be used throughout an application, and conversions performed as necessary.

The Explicit Form

The best known line equation is this one:

$$y = mx + c$$

Because the value of m becomes very large as the line comes near to vertical, this formulation is practically useless for computation, as operations based upon it would be riddled with special cases.

We have used the common notation for this equation, as it will not be mentioned again.

The Implicit Form

This is a more stable version of the line definition above:

$$ax + by + c = 0$$

As it stands, this equation can be multiplied through by any non-zero constant without altering its meaning. It is made more useful, and potential numerical problems are avoided, by putting it into *canonical* or *normalised* form by imposing the constraint:

$$a^2 + b^2 = 1$$

This is most simply achieved by multiplying through by:

$$\frac{1}{\sqrt{(a^2 + b^2)}}$$

In the normalised form a and b are *direction cosines*, the cosines of the angles which the normal to the line makes with the x and y axes. The absolute value of c is the distance from the line to the origin.

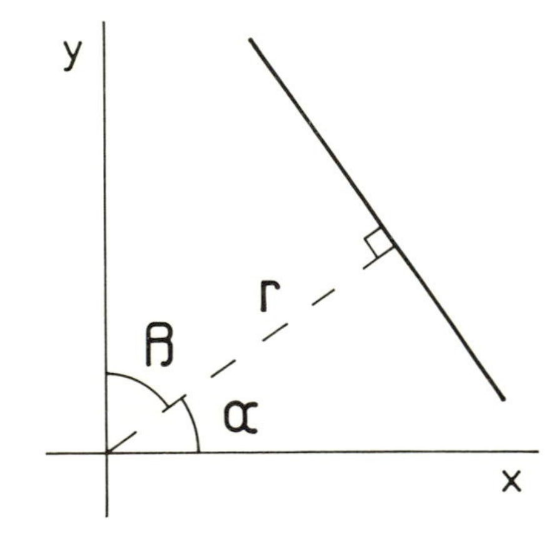

$$a = \cos\alpha\ , \quad b = \cos\beta\ , \quad c = -r$$

The entire equation may still be multiplied by −1 without violating the normalising $a^2 + b^2 = 1$ condition. We may choose to prefer neither form, or to have a convention that c is always positive (although this still leaves ambiguity when c = 0 and the line goes through the origin). Alternatively we may use the sign of c to convert the line to the boundary of a region in the plane, with an inside and an outside, or to impose a direction on the line. Where the line is a boundary, it may be called a *linear half-plane*, because it bisects the plane into two semi-infinite areas.

The notion of side may be expressed by a convention that the vector (see Section 6.1) formed from the direction cosines always points towards the outside (or the inside) of the region. Direction along the line may be specified as either right- or left-handed from the normal vector.

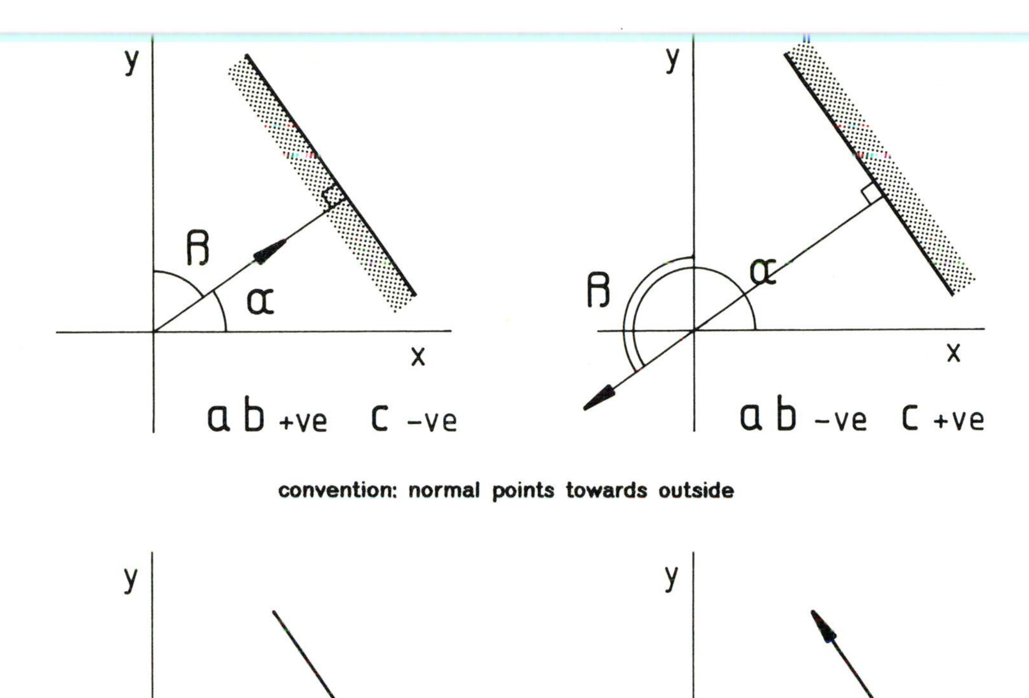

convention: normal points towards outside

convention: line direction is to right of normal vector

The Parametric Form

This form of line equation consists of two equations which give x and y in terms of a third variable, the parameter, t:

$$x = x_0 + ft$$

$$y = y_0 + gt$$

The convention x_0 and y_0 for the constant terms is adopted because it will readily be seen that (x_0, y_0) is the point on the line corresponding to a zero value of the parameter, t.

These equations have four constant terms, as opposed to three in the implicit form. The extra freedom allows us to specify just how the parameter varies along the line. There are two useful conventions. The first is that the parameter varies between 0 and 1 over a given line segment. This formulation is dealt with in the section on a line between two points (Section 1.6). The second convention (the normalised form) makes t correspond to real distance along the line.

A line through a point J, making angles α and β with the x and y axes respectively

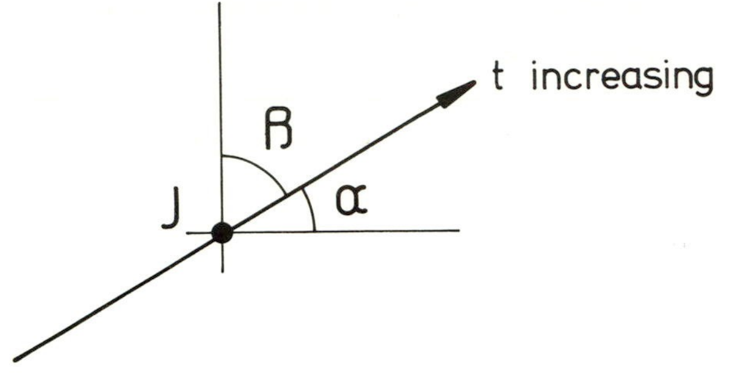

has the parametric equations:

$$x = x_J + (\cos\alpha)t$$

$$y = y_J + (\cos\beta)t$$

In general any parametric line can be normalised by dividing the coefficients of t (but *not* the constant terms) by $\sqrt{(f^2 + g^2)}$, in a similar manner to the way that the implicit form was normalised.

Conversion from Implicit to Parametric Form.

A general, not necessarily normalised, implicit line ax + by +c = 0 is conveniently parameterised as:

$$x = \frac{-ac}{(a^2 + b^2)} + bt$$

$$y = \frac{-bc}{(a^2 + b^2)} - at$$

This operation can be coded:

```
ROOT = 1.0/(A*A + B*B)
FACTOR = -C*ROOT
XO = A*FACTOR
YO = B*FACTOR
```

```
ROOT = SQRT(ROOT)
F = B*ROOT
G = -A*ROOT
```

(This assumes that the data structure is not corrupted, and that A and B are not both zero. It may be worth checking for this.)

This conversion also normalises the equation. If the implicit equation to be converted is known to be normalised already, then the division by $(a^2 + b^2)$ can be omitted.

The point (x_0, y_0) in this parametric form is the point on the line where the line meets the normal to the origin. The direction of parameterisation has t increasing in the line direction corresponding to the right handed convention in the implicit form. For instance, if a and b were positive, and c was negative, the parameter would behave as shown below:

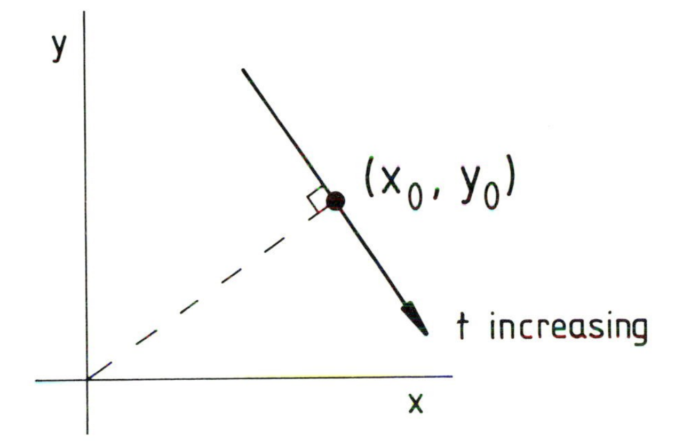

Conversion from Parametric to Implicit Form

A parametric line

$$x = x_0 + ft$$

$$y = y_0 + gt$$

can readily be converted to the implicit form:

$$-gx + fy + (x_0 g - y_0 f) = 0$$

which is only normalised if the parametric form was also normalised. The increasing t direction of the parametric line will again be to the right of the normal vector from the origin to the line.

1.3 Distance from a Point to a Line

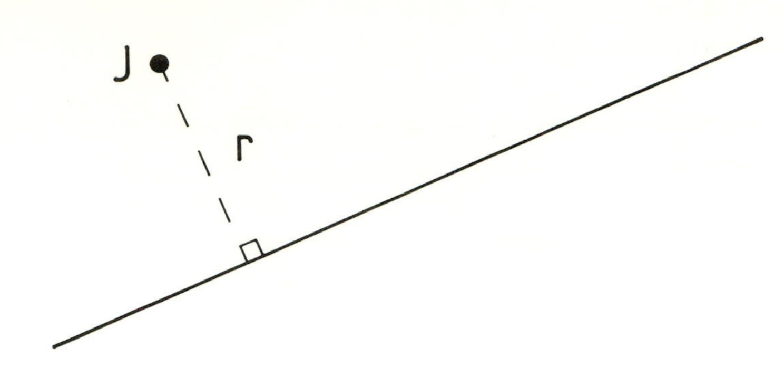

If the line is in its implicit form and has the equation

$$ax + by + c = 0$$

and the point is (x_J, y_J), then the shortest distance of that point from the line, r, is the length of a perpendicular drawn from the point to the line. r^2 is given by:

```
ABSQ = A*A + B*B
IF (ABSQ.LT.ACCY) THEN

        ..... The line is improperly defined

ELSE
        SR = A*XJ + B*YJ + C
        RSQ = SR*SR/ABSQ
ENDIF
```

If the equation of the line is in its normalised form, ABSQ will be 1.0 and this code can be simplified to the single statement

```
SR = A*XJ + B*YJ + C
```

The sign of SR indicates on which side of the line (x_J, y_J) lies. Positive values indicate that the point is on the side of the line in the direction that the vector (a,b) is pointing (see Section 6.1). Negative values indicate that it is on the other side. If this information is irrelevant, the function ABS(SR) should be calculated.

If the line is in parametric form

$$x = x_0 + ft$$

$$y = y_0 + gt$$

the code becomes a little more complicated:

```
FSQ = F*F
GSQ = G*G
FGSQ = FSQ + GSQ
IF (FGSQ.LT.ACCY) THEN

        ..... The line is improperly defined

ELSE
        XJ0 = XJ - X0
        YJ0 = YJ - Y0
        FG = F*G
        FINV = 1.0/FGSQ
        DX = GSQ*XJ0 - FG*YJ0
        DY = FSQ*YJ0 - FG*XJ0
        RSQ = (DX*DX + DY*DY)*FINV*FINV
ENDIF
```

The value of the parameter, t, at the closest point on the line to (x_J, y_J) is given by:

```
TJ = (F*XJ0 + G*YJ0)*FINV
```

1.4 Angle between Two Lines

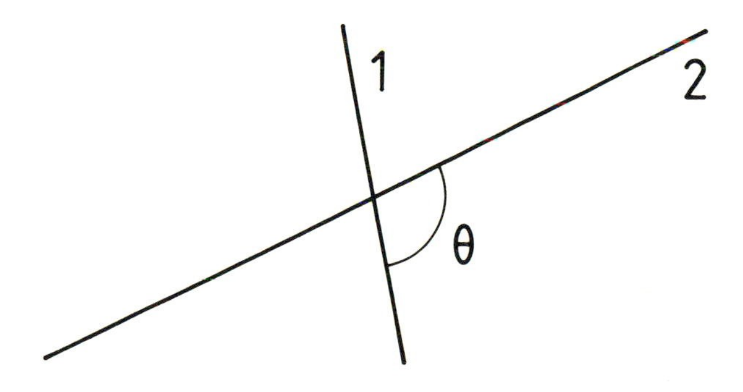

The angle between two lines is found from their direction cosines. If the two line equations are normalised and are

$$a_1x + b_1y + c_1 = 0$$

$$a_2x + b_2y + c_2 = 0$$

then the angle between them is

$$\theta = \cos^{-1}(a_1a_2 + b_1b_2)$$

and for two normalised parametric lines

$$x = x_1 + f_1s$$

$$y = y_1 + g_1s$$

$$x = x_2 + f_2t$$

$$y = y_2 + g_2t$$

the angle is:

$$\theta = \cos^{-1}(f_1f_2 + g_1g_2)$$

If the lines are not normalised then, rather than normalising each separately for just this one operation, it is quicker to use the expressions:

$$\theta = \cos^{-1}\frac{a_1a_2 + b_1b_2}{\sqrt{[(a_1^2 + b_1^2)(a_2^2 + b_2^2)]}}$$

(implicit)

and $$\theta = \cos^{-1}\frac{f_1f_2 + g_1g_2}{\sqrt{[(f_1^2 + g_1^2)(f_2^2 + g_2^2)]}}$$

(parametric)

These use one square root each, instead of the two required to normalise each equation separately, and, in the case of the implicit form, the unnecessary normalisation of the constant term is also avoided.

The FORTRAN ACOS (arc cosine) function returns angles in the range 0 to π. If the acute angle between the lines is required, then values of θ that are greater than $\pi/2$ must be subtracted from π. Alternatively, if the lines have a direction associated with them, or are half-planes, the value of θ is the angle through which one line would have to be turned in order to correspond exactly with the other.

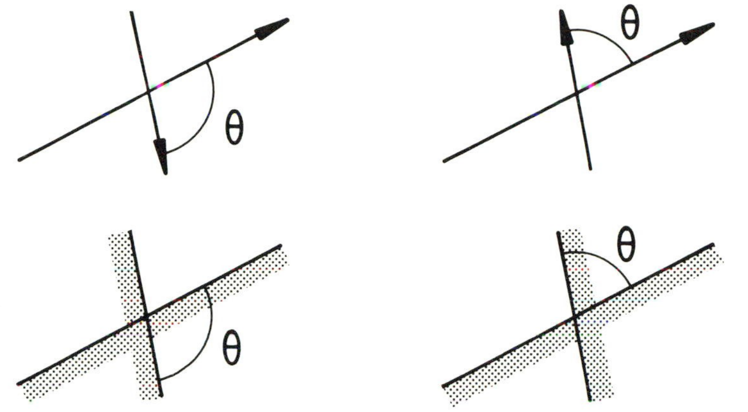

Some elderly FORTRAN compilers have no ACOS function. An ACOS function can be written in terms of the ubiquitous ATAN2 function.

```
FUNCTION ACOS(X)
     ACOS = ATAN2(SQRT(1.0 - X*X), X)
RETURN
END
```

Both this function and the generic ACOS will give an error if the argument is greater than 1.0. In the function above this will take the form of an attempt to take the square root of a negative number. The generic ACOS routine will also give an error if its argument is less than zero, whereas the code above will return a negative angle for arguments between -1.0 and 0.0, and only give an error for arguments less than -1.0.

In any case, it is best to ensure that the argument is in the range 0.0 to 1.0, as numerical errors may produce values just outside this range even if the original data were correct. The code:

```
THETA = ACOS(AMIN1(1.0,AMAX1(0.0,X)))
```

solves this problem, but gross errors arising from corrupted data must be detected separately.

If it is required only to find pairs of lines, the angles between which fall within a particular range of values, it will be more efficient to take the cosines of the limiting values and compare these with the products of the direction cosines directly. This avoids many calls to the costly ACOS function. For instance, lines within 1° of being parallel will have a sum of products of direction cosines within the range -1.0 to -0.99985 [cos(179°)] or 0.99985 to 1.0. Lines within 1° of perpendicularity will have values in the range -0.017453 [cos(91°)] and 0.017453 [cos(89°)].

This is an appropriate place to note that we recommend that angles are avoided wherever possible, because of the cost of trigonometric functions and the potential numerical instabilities that they introduce. However, especially when angles are required for input or output, the use of trigonometric functions cannot be avoided. It can be minimised by using the following relations between sines, cosines, tangents, and tangents of half-angles (used in describing arcs).

$$\sin\theta = \sqrt{(1 - \cos^2\theta)} = \frac{\tan\theta}{\sqrt{(1 + \tan^2\theta)}} = \frac{2\tan(\theta/2)}{1 + \tan^2(\theta/2)}$$

$$\cos\theta = \sqrt{(1 - \sin^2\theta)} = \frac{1}{\sqrt{(1 + \tan^2\theta)}} = \frac{1 - \tan^2(\theta/2)}{1 + \tan^2(\theta/2)}$$

$$\tan\theta = \frac{\sin\theta}{\sqrt{(1 - \sin^2\theta)}} = \frac{\sqrt{(1 - \cos^2\theta)}}{\cos\theta} = \frac{2\tan(\theta/2)}{1 - \tan^2(\theta/2)}$$

$$\tan(\theta/2) = \frac{1 - \sqrt{(1 - \sin^2\theta)}}{\sin\theta} = \sqrt{\frac{1 - \cos\theta}{1 + \cos\theta}} = \frac{\sqrt{(1 + \tan^2\theta)} - 1}{\tan\theta}$$

1.5 Intersection of Two Lines

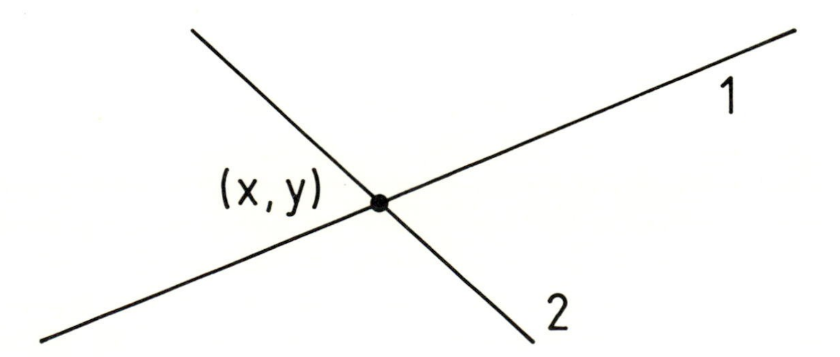

This problem is one well known to numerical analysts - the solution of two linear simultaneous equations. If the equations of the lines are both in implicit form

$$a_1x + b_1y + c_1 = 0$$

$$a_2x + b_2y + c_2 = 0$$

and they intersect at the point (x, y) then the solution is simply coded:

```
DET = A1*B2 - A2*B1
IF (ABS(DET).LT.ACCY) THEN

        ..... The two lines are parallel

ELSE
        DINV = 1.0/DET
        X = (B1*C2 - B2*C1)*DINV
        Y = (A2*C1 - A1*C2)*DINV
ENDIF
```

If one equation is in its implicit form and the other is parametric the solution is a little more complicated.

$ax + by + c = 0$ *Implicit line equation*

$x = x_0 + ft$ *Parametric*

$y = y_0 + gt$ *line equation*

This gives the code:

```
DET = A*F + B*G
IF (ABS(DET).LT.ACCY) THEN

        ..... The two lines are parallel

ELSE
        DINV = 1.0/DET
        PDET = X0*G - Y0*F
        X = (B*PDET - C*F)*DINV
        Y = -(A*PDET + C*G)*DINV
ENDIF
```

At the point of intersection:

$$t = \frac{-(c + ax_0 + by_0)}{(af + bg)}$$

If both lines are in parametric form

$$x = x_1 + f_1 s$$

$$y = y_1 + g_1 s$$

and

$$x = x_2 + f_2 t$$

$$y = y_2 + g_2 t$$

the solution becomes

```
F1G2 = F1*G2
F2G1 = F2*G1
DET = F2G1 - F1G2
IF (ABS(DET).LT.ACCY) THEN

        ..... The two lines are parallel

ELSE
        Y21 = Y2 - Y1
        X21 = X2 - X1
        DINV = 1.0/DET
        S = (F2*Y21 - G2*X21)*DINV
        T = (F1*Y21 - G1*X21)*DINV
        X = X1 + F1*S
        Y = Y1 + G1*S
ENDIF
```

where S and T are the parameter values at the point of intersection. If only the coordinates of the point are to be calculated there is no need to include the line that calculates T.

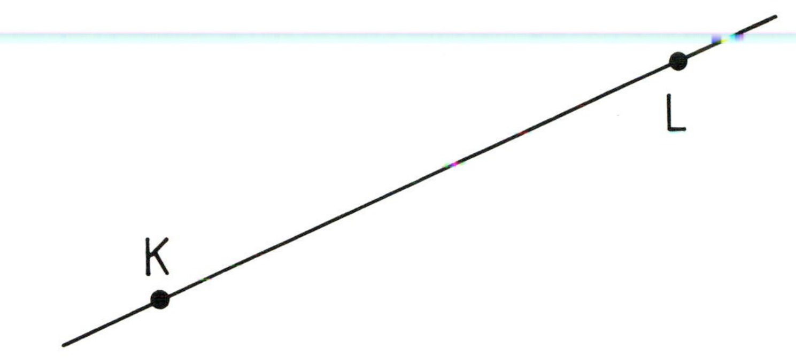

This is the second common way of specifying a parametric line (see Section 1.2). The line is specified so that t = 0 at the first point and t = 1 at the second. The equations for this form of line are:

$$x = x_K + (x_L - x_K)t$$

$$y = y_K + (y_L - y_K)t$$

This is not normalised unless the distance between K and L is 1, of course. The implicit form is found from the conversion formula given in section 1.2, and cannot be simplified beyond this point:

$$(y_K - y_L)x + (x_L - x_K)y + (x_K y_L - x_L y_K) = 0$$

In general this equation is not normalised either.

Both forms can be normalised by dividing the entire implicit equation, or the terms in t of the parametric equation, by the distance from K to L.

For example, the coefficients of a normalised implicit equation can be computed as follows:

```
XLK = XL - XK
YLK = YL - YK
RSQ = XLK*XLK + YLK*YLK
IF (RSQ.LT.ACCY) THEN

        ..... The points coincide

ELSE
        RINV = 1.0/SQRT(RSQ)
        A = -YLK*RINV
```

```
            B = XLK*RINV
            C = (XK*YL - XL*YK)*RINV
      ENDIF
```

1.7 Line Equidistant from Two Points

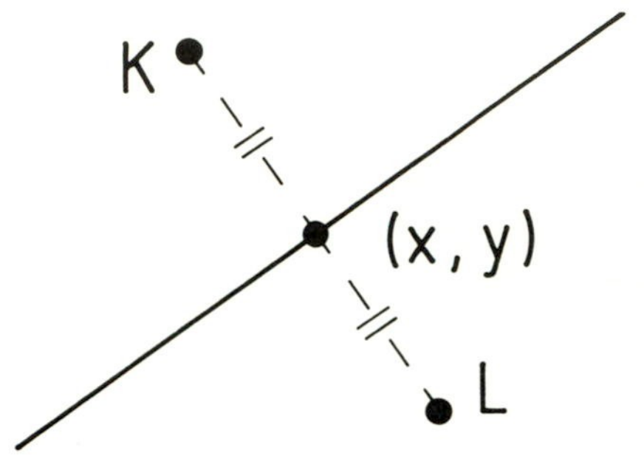

Both the implicit and the parametric forms of this line come out most simply in forms that are not normalised. The implicit equation is:

$$(x_L - x_K)x + (y_L - y_K)y - \frac{1}{2}[(x_L^2 + y_L^2) - (x_K^2 + y_K^2)] = 0$$

The normal vector to the line will always point from K to L.

The parametric form is chosen so that the t = 0 point is the mid-point of the line joining K and L.

$$x = \frac{(x_K + x_L)}{2} - (y_L - y_K)t$$

$$y = \frac{(y_K + y_L)}{2} + (x_L - x_K)t$$

The parameter, t, increases to the right of the line from K to L.

Both forms of equation are normalised by dividing by the distance from K to L: the entire implicit equation or just the terms in t of the parametric equations.

For example, the terms of a normalised parametric representation could be computed as follows:

```
XLK = XL - XK
YLK = YL - YK
RSQ = XLK*XLK + YLK*YLK
IF (RSQ.LT.ACCY) THEN

        ..... the two points coincide.

ELSE
        RINV = 1.0/SQRT(RSQ)
        XO = 0.5*(XK + XL)
        F = -YLK*RINV
        YO = 0.5*(YK + YL)
        G = XLK*RINV
ENDIF
```

1.8 Normal to a Line Through a Point

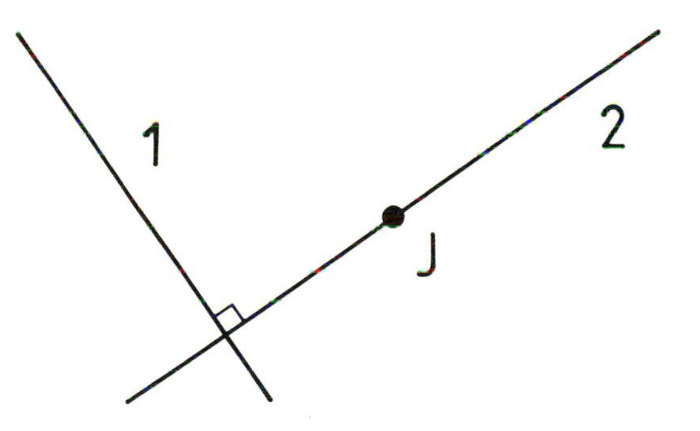

If the line equation is in implicit form

$$a_1x + b_1y + c_1 = 0$$

and the point is (x_J, y_J), and the equation of a line at right angles to the first line passing through that point that we want to find will be

$$a_2x + b_2y + c_2 = 0$$

then the terms of that second equation are:

$$a_2 = b_1$$

$$b_2 = -a_1$$

$$c_2 = a_1 y_J - b_1 x_J$$

Note that if the original line equation was in its normalised form this property is transferred to the second line.

If the line is in parametric form

$$x = x_1 + f_1 t$$

$$y = y_1 + g_1 t$$

then the normal to it is the line

$$x = x_J - g_1 s$$

$$y = y_J + f_1 s$$

2

Points, lines and circles

2.1 Equations of a Circle

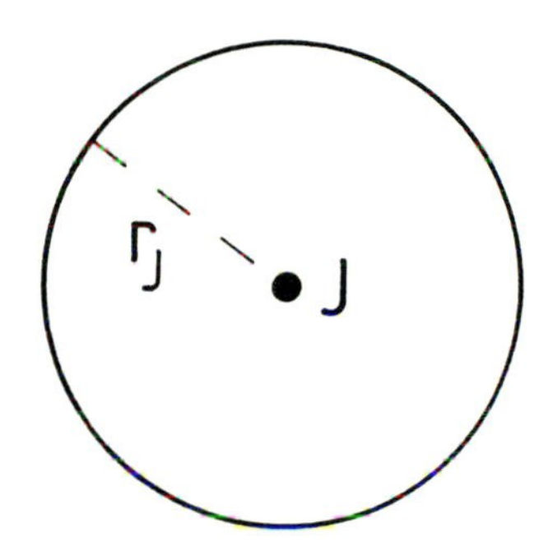

The implicit equation of a circle

$$(x - x_J)^2 + (y - y_J)^2 - r_J^2 = 0$$

is the most commonly used for whole circles. The parametric form

$$x = x_J + r_J\cos\theta$$

$$y = y_J + r_J\sin\theta$$

is also straightforward, giving a parameterisation in terms of the angle subtended at the circle centre:

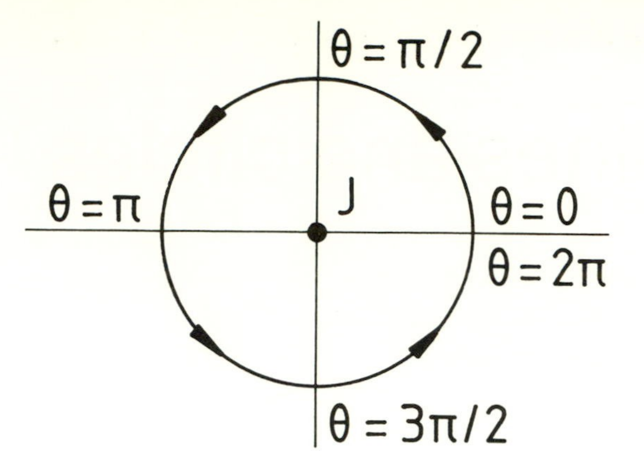

This form is particularly useful when the circle is the path of a rotating object, as the parameter, θ, can describe any number of rotations.

Alternatively, the parameterisation in terms of the half-angle subtended at the centre:

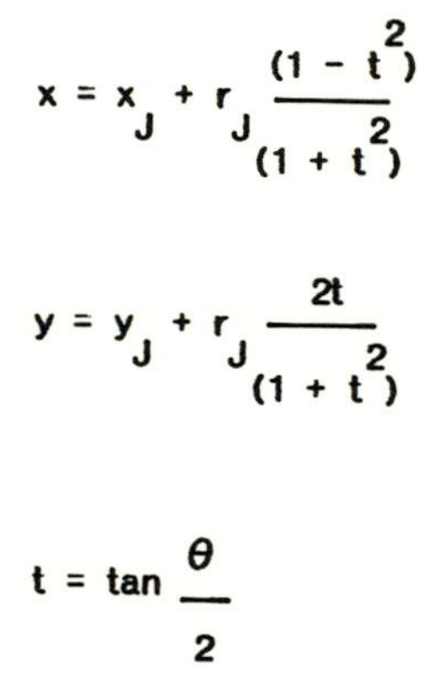

$$x = x_J + r_J \frac{(1 - t^2)}{(1 + t^2)}$$

$$y = y_J + r_J \frac{2t}{(1 + t^2)}$$

$$t = \tan \frac{\theta}{2}$$

eliminates the need for trigonometric functions. It defines the circle as follows:

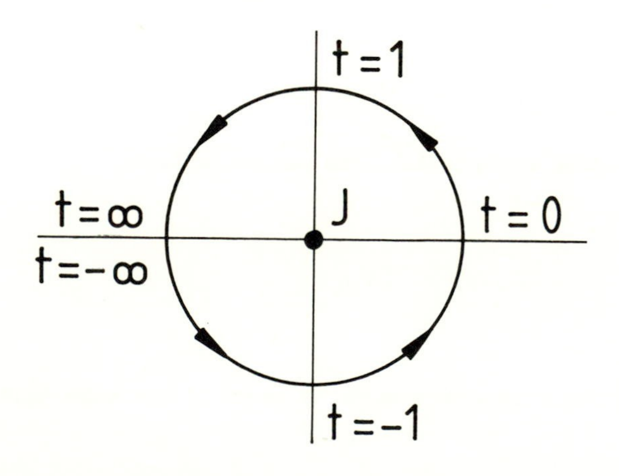

but is normally only used to define a single quadrant, $0 < t < 1$, to avoid the obvious numerical problems.

Both parametric forms are useful for describing arcs; Section 3.4 gives details of this.

2.2 Intersections of a Line and a Circle

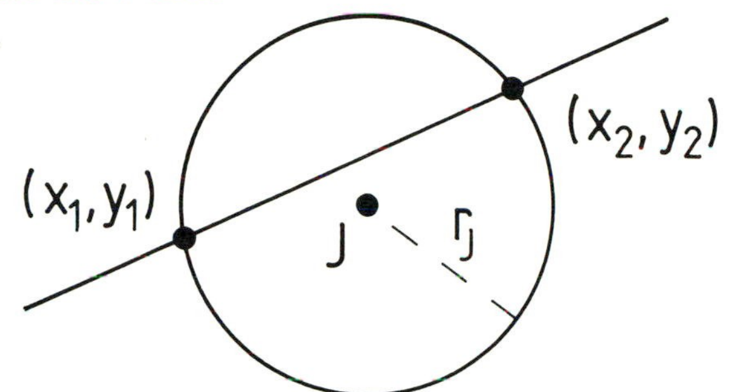

This problem is most easily solved if the circle is in implicit form

$$(x - x_J)^2 + (y - y_J)^2 - r_J^2 = 0$$

and the line is parametric:

$$x = x_0 + ft$$

$$y = y_0 + gt$$

Substituting the parametric equations into the circle equation gives a quadratic in t, the two roots of which represent the points on the line where it cuts the circle. If the roots are imaginary, then the line does not cut the circle at all. If the roots are coincident the line is tangential to the circle.

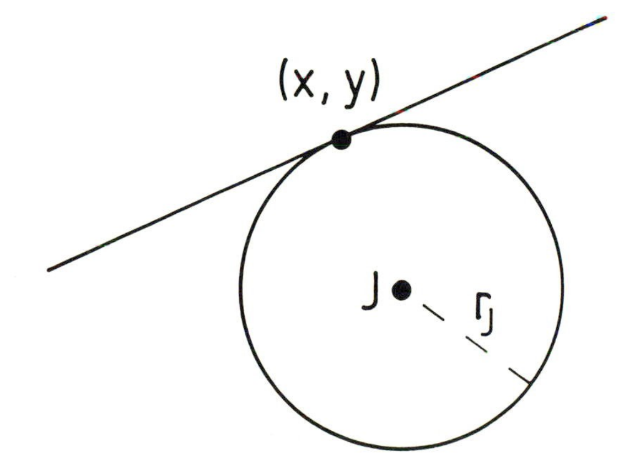

The value of t at the intersection points is

$$t = \frac{f(x_J - x_0) + g(y_J - y_0) \pm \sqrt{\{r_J^2(f^2 + g^2) - [f(y_0 - y_J) - g(x_0 - x_J)]^2\}}}{(f^2 + g^2)}$$

and the points are found by substituting these values back into the parametric equations. This is coded:

```
FSQ = F*F
GSQ = G*G
FGSQ = FSQ + GSQ
IF (FGSQ.LT.ACCY) THEN

        ..... Line coefficients are corrupt

ELSE
        XJO = XJ - XO
        YJO = YJ - YO
        FYGX = F*YJO - G*XJO
        ROOT = RJ*RJ*FGSQ - FYGX*FYGX
        IF (ROOT.LT.-ACCY) THEN

                ..... Line does not intersect circle

        ELSE
                FXGY = F*XJO + G*YJO
                IF (ROOT.LT.ACCY) THEN
                        T = FXGY/FGSQ           Line is tangential
                        X = XO + F*T
                        Y = YO + G*T
                ELSE
                        ROOT = SQRT(ROOT)       Two intersections
                        FGINV = 1.0/FGSQ
                        T1 = (FXGY - ROOT)*FGINV
                        T2 = (FXGY + ROOT)*FGINV
                        X1 = XO + F*T1
                        Y1 = YO + G*T1
                        X2 = XO + F*T2
                        Y2 = YO + G*T2
                ENDIF
        ENDIF
ENDIF
```

Note that, if the parametric line is normalised, the variable FGSQ will be 1.0, and the code can be simplified. To convert an implicit line equation to parametric form use the method described in Section 1.2

2.3 Intersections of Two Circles

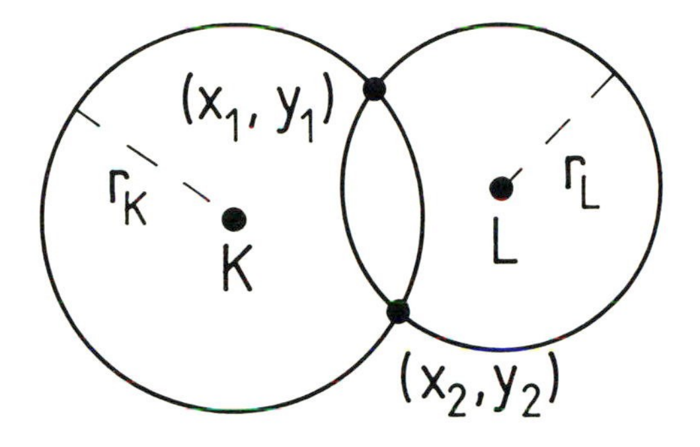

Two circles may have two intersection points or one intersection point at a common tangent.

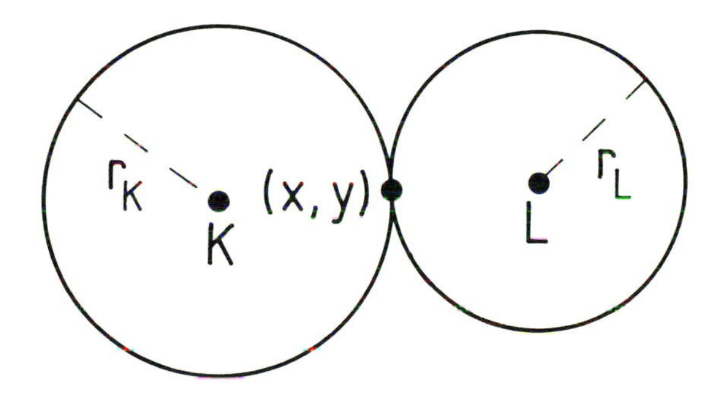

Alternatively they may not intersect at all. The position of the intersection points may be found by applying Pythagoras' theorem, which will give the parametric equation of the line on which the intersection points lie, and then solving the resulting quadratic equation in the parameter as was done in the last section. The substitution back into the parametric line equation can be done at the same time to shorten the code, which mirrors the algebra:

```
RKSQ = RK*RK          It is more efficient to use squared radii
RLSQ = RL*RL          and to omit this
XLK = XL - XK
YLK = YL - YK

DISTSQ = XLK*XLK + YLK*YLK
```

```
IF (DISTSQ.LT.ACCY) THEN

          ..... The two circles have the same centre

ELSE
          DELRSQ = RLSQ - RKSQ
          SUMRSQ = RKSQ + RLSQ
          ROOT = 2.0*SUMRSQ*DISTSQ-DISTSQ*DISTSQ-DELRSQ*DELRSQ
          IF (ROOT.LT.-ACCY) THEN

                    ..... The circles do not intersect

          ELSE
                    DSTINV = 0.5/DISTSQ
                    SCL = 0.5 - DELRSQ*DSTINV
                    X = XLK*SCL + XK
                    Y = YLK*SCL + YK
                    IF (ROOT.LT.ACCY) THEN

                              ..... Circles just touch at (X, Y)

                    ELSE
                              ROOT = DSTINV*SQRT(ROOT) Two
                              XFAC = XLK*ROOT          intersections
                              YFAC = YLK*ROOT
                              X1 = X - YFAC
                              Y1 = Y + XFAC
                              X2 = X + YFAC
                              Y2 = Y - XFAC
                    ENDIF
          ENDIF
ENDIF
```

The straight line between the points of intersection and, in the limit, the common tangent is given by the implicit equation

$$ax + by + c = 0$$

where

$$a = x_L - x_K$$

$$b = y_L - y_K$$

and $$c = \frac{[(r_L^2 - r_K^2) - (x_L - x_K)^2 - (y_L - y_K)^2]}{2} - x_K(x_L - x_K) - y_K(y_L - y_K)$$

This equation is not normalised.

2.4 Tangents from a Point to a Circle

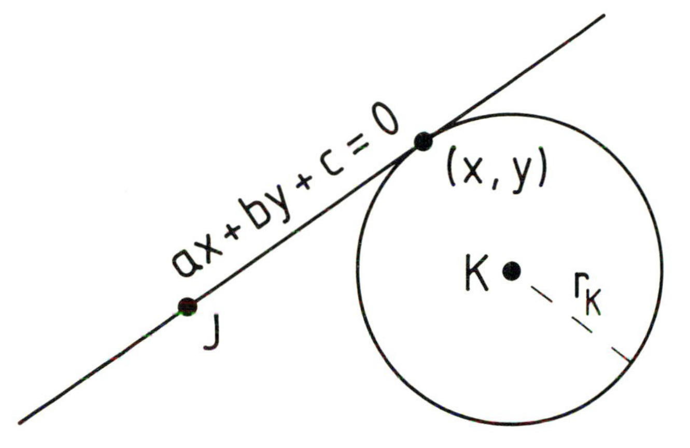

If the point is outside the circle there are two tangents to it, if it is just on the circumference there is one, and if it is inside there is none.

If the equation of the tangent line required is

$$ax + by + c = 0$$

Then the coefficients a and b are obtained from:

$$a = \frac{\mp r_K(x_K - x_J) - (y_K - y_J)\sqrt{[(x_K - x_J)^2 + (y_K - y_J)^2 - r_K^2]}}{(x_K - x_J)^2 + (y_K - y_J)^2}$$

$$b = \frac{\mp r_K(y_K - y_J) + (x_K - x_J)\sqrt{[(x_K - x_J)^2 + (y_K - y_J)^2 - r_K^2]}}{(x_K - x_J)^2 + (y_K - y_J)^2}$$

and c can then be calculated from the fact that the tangent passes through J:

$$c = -ax_J - by_J$$

There are normally two possible tangent lines, obtained by attaching a sign to r_K. If the square root is positive then the two tangents are obtained as follows:

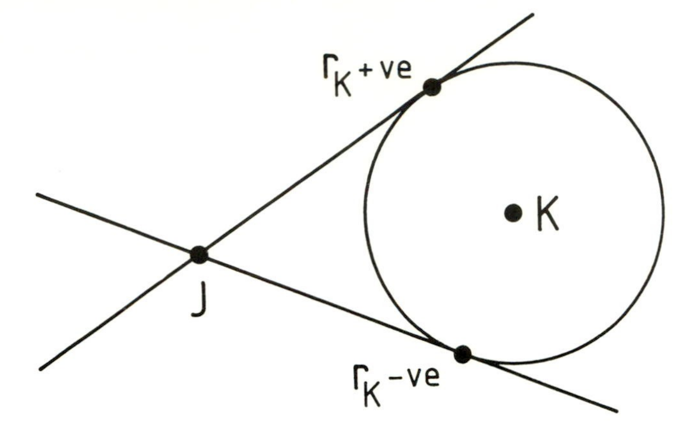

If the point J lies on the circle the contents of the square root are zero, and there is only a single tangent. If J is inside the circle there are no tangents, of course, and the root goes negative. Repeated factors make the corresponding code short:

```
XKJ = XK - XJ
YKJ = YK - YJ

XKJSQ = XKJ*XKJ
YKJSQ = YKJ*YKJ
DENOM = XKJSQ + YKJSQ
IF (DENOM.LT.ACCY) THEN

        ..... J and K are coincident

ELSE
        ROOT = DENOM - RK*RK
        IF (ROOT.LT.-ACCY) THEN

                ..... J is within the circle

        ELSE
                DENINV = 1.0/DENOM
                IF (ROOT.LT.ACCY) THEN
                        A = -RK*XKJ*DENINV          J lies on circle
                        B = -RK*YKJ*DENINV
                ELSE
```

```
                    ROOT = SQRT(ROOT)
                    RKSIGN = RK                 Negate RKSIGN
                                                for other
                                                tangent

                    A = (-YKJ*ROOT - RKSIGN*XKJ)*DENINV
                    B = (XKJ*ROOT - RKSIGN*YKJ)*DENINV
                ENDIF
                C = -(A*XJ + B*YJ)
            ENDIF
        ENDIF
```

If the coordinates of the tangent point are required, they can be obtained from the a and b coefficients of the appropriate line:

$$x = x_K + ar_K$$

$$y = y_K + br_K$$

2.5 Tangents to a Circle Normal to a Line

This problem always has two solution lines, one each side of the circle.

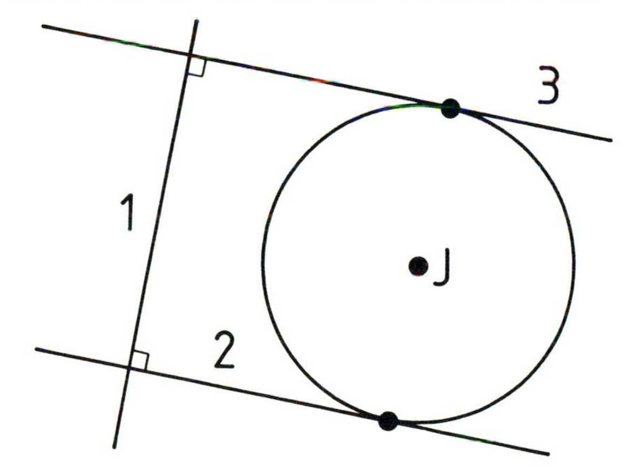

If the known line is

$$a_1x + b_1y + c_1 = 0$$

and the circle is

$$(x - x_J)^2 + (y - y_J)^2 - r_J^2 = 0$$

then the two new lines tangential to the circle and normal to the given line will be

$$a_{2,3}x + b_{2,3}y + c_{2,3} = 0$$

where

$$a_{2,3} = \mp \frac{b_1}{\sqrt{(a_1^2 + b_1^2)}}$$

$$b_{2,3} = \pm \frac{a_1}{\sqrt{(a_1^2 + b_1^2)}}$$

and $$c_{2,3} = r_J - a_{2,3}x_J - b_{2,3}y_J$$

which can be coded:

```
ROOT = A1*A1 + B1*B1
IF (ROOT.LT.ACCY) THEN

          ..... The line equation is corrupt

ELSE
          DENOM = 1.0/SQRT(ROOT)
          AFAC = B1*DENOM
          BFAC = A1*DENOM
          A2 = AFAC
          B2 = -BFAC
          C2 = RJ - A2*XJ - B2*YJ
          A3 = -AFAC
          B3 = BFAC
          C3 = RJ - A3*XJ - B3*YJ
ENDIF
```

If the vector (a,b) points towards J from the given line, then lines 2 and 3 are as labelled in the diagram. Otherwise they are reversed. The tangent points are given by:

$$x_{2,3} = x_J + a_{2,3}r_J$$

$$y_{2,3} = y_J + b_{2,3}r_J$$

Note that the lines produced will be normalised, and that if the given line is normalised the code simplifies because ROOT and DENOM both become 1.0.

2.6 Tangents between Two Circles

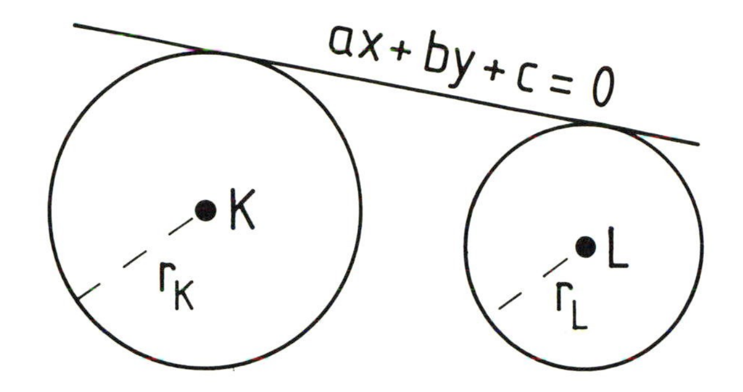

If the tangent between the circles is

$$ax + by + c = 0$$

then its coefficients are given by:

$$a = \frac{(\mp r_L \pm r_K)(x_L - x_K) - (y_L - y_K)\sqrt{[(x_L - x_K)^2 + (y_L - y_K)^2 - (\pm r_L \mp r_K)^2]}}{(x_L - x_K)^2 + (y_L - y_K)^2}$$

$$b = \frac{(\mp r_L \pm r_K)(y_L - y_K) + (x_L - x_K)\sqrt{[(x_L - x_K)^2 + (y_L - y_K)^2 - (\pm r_L \mp r_K)^2]}}{(x_L - x_K)^2 + (y_L - y_K)^2}$$

$$c = \mp r_K - ax_K - by_K$$

The signs attached to the two circle radii determine which of the four possible tangents

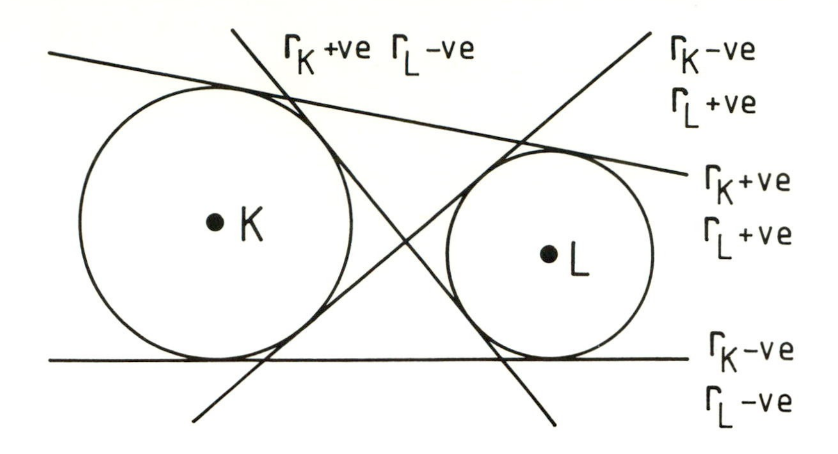

is to be found.

If the two circles intersect, only two tangents are possible:

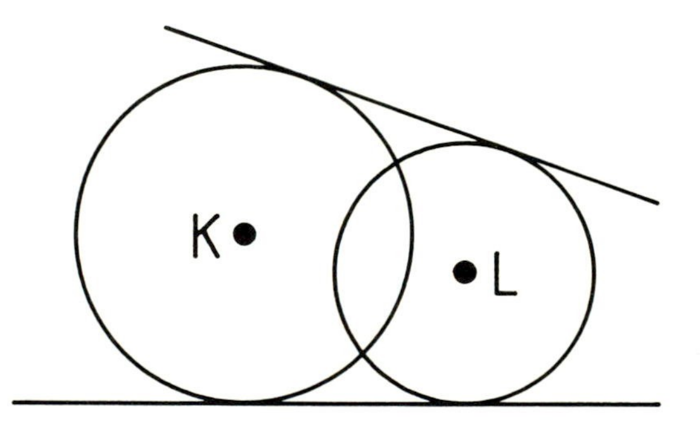

Attempting to calculate either of the non-existant tangents will lead to a negative expression to be square rooted in the formula.

Finding a tangent may be coded:

```
RLK = RL - RK
```

Assumes both circles to be touched anticlockwise. RK and/or RL must be negated otherwise

```
YLK = YL - YK
XLK = XL - XK

XLKSQ = XLK*XLK
YLKSQ = YLK*YLK
```

```
      DENOM = XLKSQ + YLKSQ
      IF (DENOM.LT.ACCY) THEN

            ..... Circle centres are coincident

      ELSE
            ROOT = DENOM - RLK*RLK
            IF (ROOT.LT.-ACCY) THEN

                  ..... This tangent does not exist

            ELSE
                  ROOT = SQRT(AMAX1(0.0,ROOT))
                  DENINV = 1.0/DENOM
                  A = (-RLK*XLK - YLK*ROOT)*DENINV
                  B = (-RLK*YLK + YLK*ROOT)*DENINV
                  C = -(RK + A*XK + B*YK)
            ENDIF
      ENDIF
```

The line equation produced will be normalised.

2.7 Circles of Given Radius through Two Points

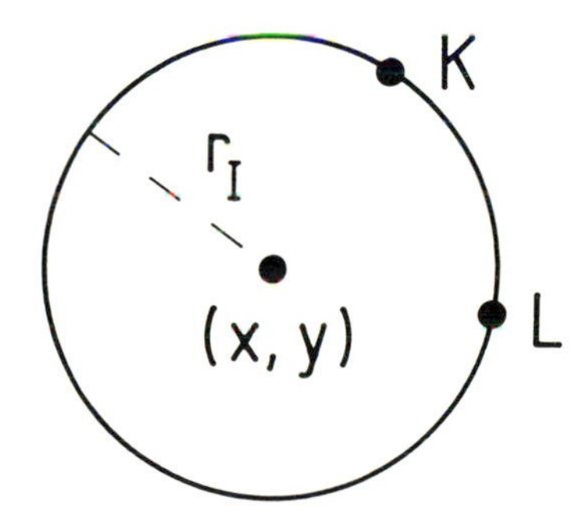

This is actually the same problem as finding the intersections of two circles with the same radius centred on the given points. These two intersection points will be the centres.

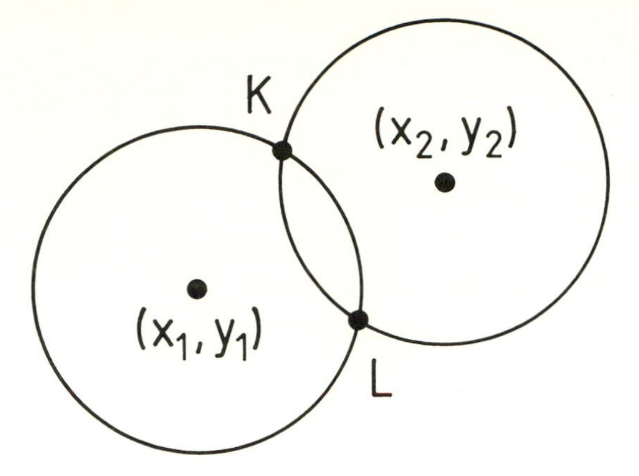

The same code can be used as that given in Section 2.3, with the added simplification that the value of the variable DELRSQ must be zero.

2.8 Circles of Given Radius through a Point and Tangent to a Line

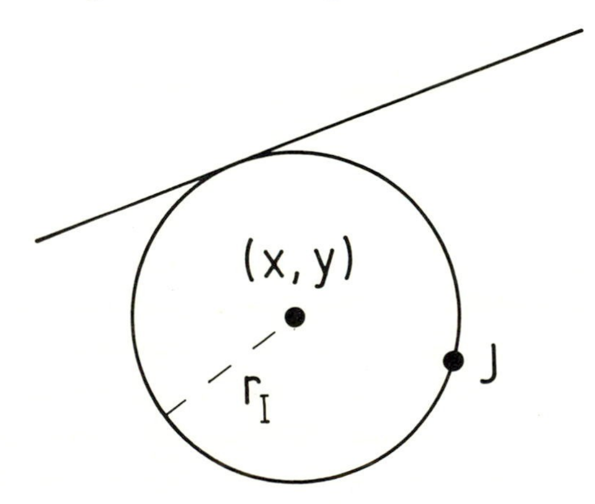

This problem is really a special case of the one dealt with later in Section 2.11. The point J is first taken as a local origin. The line equation is referred to this origin by transforming the constant term, c:

$$c' = c + ax_J + by_J$$

The value of c' must be made positive, by multiplying the whole transformed equation by -1 if necessary. The circle centre coordinates are then found from the expressions

$$x = \frac{-a(c' - r_I) \pm b\sqrt{[r_I^2(a^2 + b^2) - (c'-r_I)^2]}}{a^2 + b^2}$$

$$y = \frac{-b(c' - r_1) \mp a\sqrt{[r_1^2(a^2 + b^2) - (c'-r_1)^2]}}{a^2 + b^2}$$

which simplify to

$$x = -a(c' - r_1) \pm b\sqrt{[c'(2r_1 - c')]}$$

$$y = -b(c' - r_1) \mp a\sqrt{[c'(2r_1 - c')]}$$

if the line is normalised. The two signs from the root give the two cases on one side of the line:

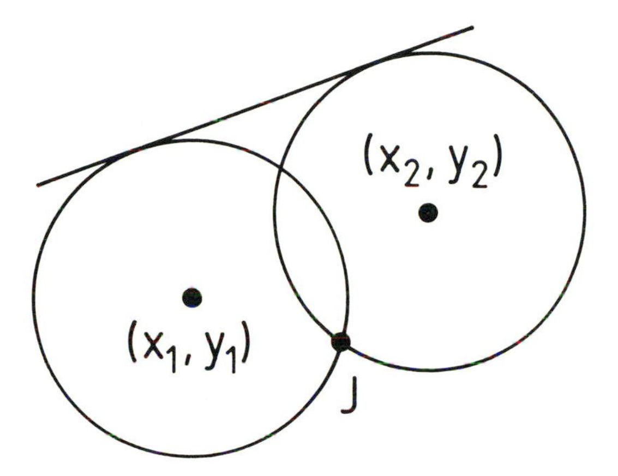

The root corresponding to the top signs (+ for x, − for y) gives the centre (x_1, y_1) to the left of the perpendicular from J to the line. The other root generates the centre (x_2, y_2) to the right of the perpendicular. A zero root indicates that there is only one possible circle, and a negative value indicates that the point is too far from the line for a circle to be created with the radius given. Lastly, note that if J is on the line, only one centre will be generated on the side of the line to which the vector (a,b) is pointing. This case is detected in the following code, *which assumes that the line is normalised.* For details of how to normalise a line equation see Section 1.2.

```
      CDASH = C + A*XJ + B*YJ
      IF (ABS(CDASH).LT.ACCY) THEN

            ..... Point J lies on the line.

      ELSE
            IF (CDASH.LT.0.0) THEN
                  ATEMP = -A
                  BTEMP = -B
                  CDASH = -CDASH
            ELSE
```

```
            ATEMP = A
            BTEMP = B
        ENDIF
        CFAC = CDASH - RI
        ROOT = RI*RI - CFAC*CFAC
        IF (ROOT.LT.-ACCY) THEN
```

..... *Point J is too far from the line.*

```
        ELSE
            IF (ROOT.LT.ACCY) THEN
                X = XJ - ATEMP*CFAC          One possible
                Y = YJ - BTEMP*CFAC           circle
            ELSE
                ROOT = SQRT(ROOT)            Two possible
                XCONST = XJ - ATEMP*CFAC     circles
                YCONST = YJ - BTEMP*CFAC
                XVAR = BTEMP*ROOT
                YVAR = ATEMP*ROOT
                X1 = XCONST + XVAR
                Y1 = YCONST - YVAR
                X2 = XCONST - XVAR
                Y2 = YCONST + YVAR
            ENDIF
        ENDIF
    ENDIF
```

29 Circles of Given Radius Tangent to Two Lines

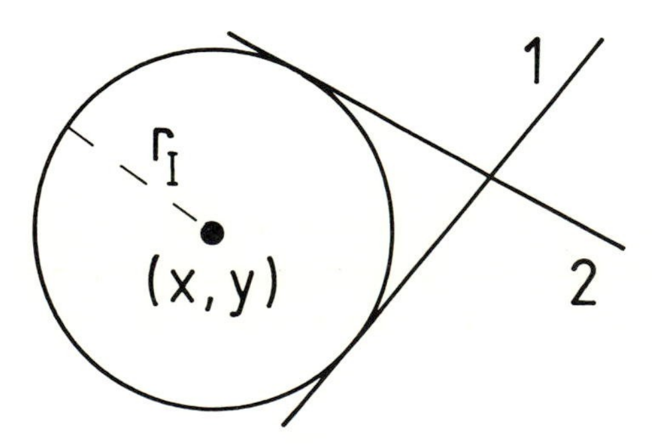

As long as the lines are not parallel (when there is either no solution, or, if the gap between the lines is the circle diameter, an infinite number of solutions) then there are four centres for the circle of given radius that make it tangential to both the lines. These centres are distributed symmetrically about the point where the lines intersect.

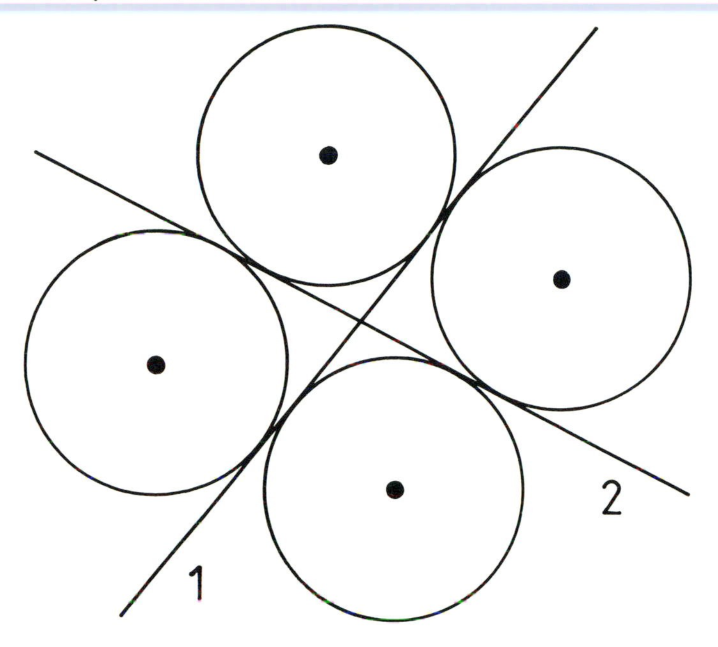

If the two lines are

$$a_1 x + b_1 y + c_1 = 0$$

and $$a_2 x + b_2 y + c_2 = 0$$

and the given radius is r_l, then the circle centres are at:

$$x = \frac{b_2[c_1 \pm r_l \sqrt{(a_2^2 + b_2^2)}] - b_1[c_2 \pm r_l \sqrt{(a_1^2 + b_1^2)}]}{(a_2 b_1 - a_1 b_2)}$$

$$y = \frac{a_2[c_1 \pm r_l \sqrt{(a_1^2 + b_1^2)}] - a_1[c_2 \pm r_l \sqrt{(a_2^2 + b_2^2)}]}{(a_1 b_2 - a_2 b_1)}$$

Note that the denominator for y is minus the denominator for x. If the first r_l in each equation is made negative, the centre will be on the side of the first line corresponding to the vector (a_1, b_1); if positive it will be on the opposite side. The second r_l terms affect the centre position similarly with respect to the second line.

This is coded:

```
DETERM = A2*B1 - A1*B2
IF (ABS(DETERM).LT.ACCY) THEN

        ..... The lines are parallel

ELSE
        AB1 = SQRT(A1*A1 + B1*B1)
        AB2 = SQRT(A2*A2 + B2*B2)

        C1RAB1 = C1 + RI*AB1           To get the four solutions
        C2RAB2 = C2 + RI*AB2           change the + here to -

        DETINV = 1.0/DETERM
        X = (B2*C1RAB1 - B1*C2RAB2)*DETINV
        Y = (A1*C2RAB2 - A2*C1RAB1)*DETINV
ENDIF
```

2.10 Circles of Given Radius through a Point and Tangent to a Circle

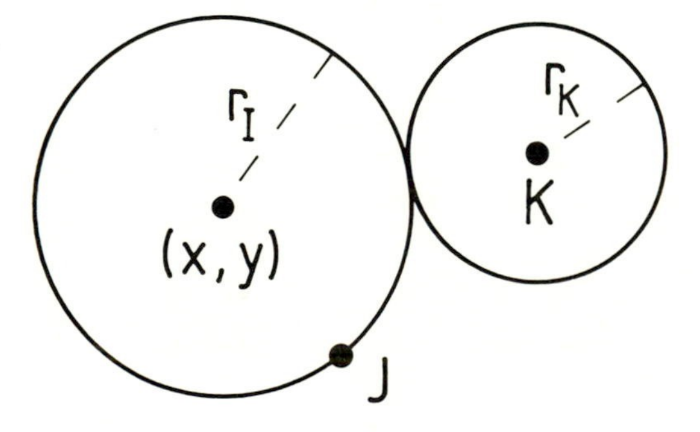

One of the given points (circle centre, K, or J) is made a local origin for the calculation; we have chosen the point J. The circle centre coordinates may then be found from

$$x = x_J + \frac{x_{KJ}[(x_{KJ}^2 + y_{KJ}^2) - r_K(2r_I + r_K)] \pm y_{KJ}s}{2(x_{KJ}^2 + y_{KJ}^2)}$$

$$y = y_J + \frac{y_{KJ}[(x_{KJ}^2 + y_{KJ}^2) - r_K(2r_I + r_K)] \mp x_{KJ}s}{2(x_{KJ}^2 + y_{KJ}^2)}$$

where

$$x_{KJ} = x_K - x_J$$

$$y_{KJ} = y_K - y_J$$

and $$s = 4r_I^2(x_{KJ}^2 + y_{KJ}^2) - [(x_{KJ}^2 + y_{KJ}^2) - r_K(2r_I + r_K)]^2$$

The sign of r_K is important. If it is positive the two circles are outside each other, as in the diagram above. If r_K is negative, one is inside the other.

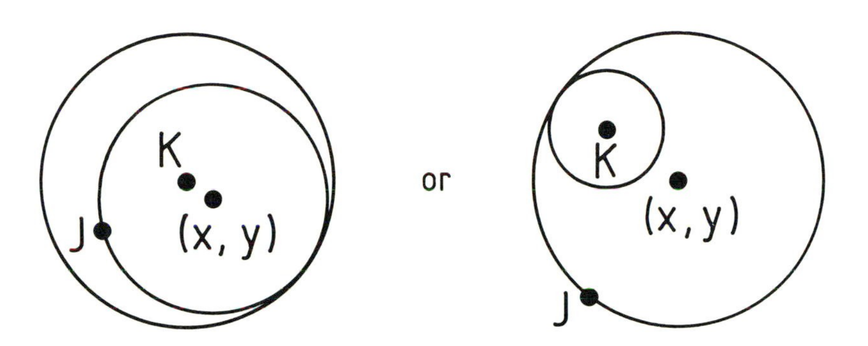

Positive and negative roots correspond to the two possible cases

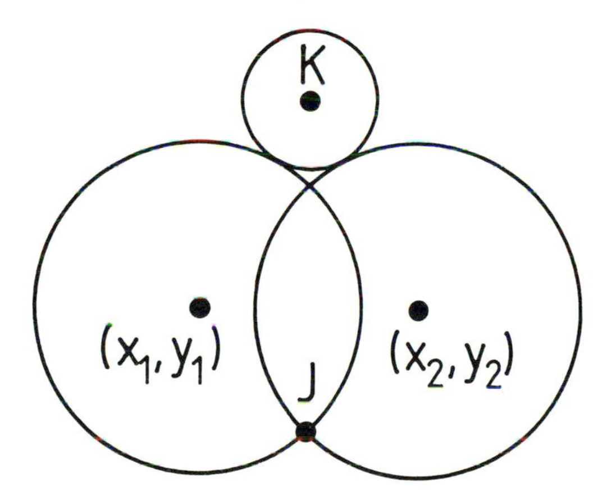

Imaginary roots indicate that the circle is required to be outside the given circle and the given point is inside, or vice versa. A zero denominator indicates that the given point and the given circle centre are coincident. The coding is economical because of the many repeated sub-expressions.

```
XKJ = XK - XJ
YKJ = YK - YJ

SQSUM = XKJ*XKJ + YKJ*YKJ
IF (SQSUM.LT.ACCY) THEN
```

```
                ..... J and K are coincident

        ELSE
                SQINV = 0.5/SQSUM
                RADSUM = (RI + RI + RK)*RK
                SUBEXP = SQSUM - RADSUM
                ROOT = 4.0*RI*RI*SQSUM - SUBEXP*SUBEXP
                SUBEXP = SUBEXP*SQINV
                IF (ROOT.LT.-ACCY) THEN

                        ..... No centre possible

                ELSE
                        IF (ROOT.LT.ACCY) THEN
                                X = XJ + XKJ*SUBEXP     Only one circle
                                Y = YJ + YKJ*SUBEXP     possible
                        ELSE
                                ROOT = SQRT(ROOT)*SQINV
                                XCONST = XJ + XKJ*SUBEXP
                                YCONST = YJ + YKJ*SUBEXP
                                XVAR = YKJ*ROOT
                                YVAR = XKJ*ROOT
                                X1 = XCONST - XVAR
                                Y1 = YCONST + YVAR
                                X2 = XCONST + XVAR
                                Y2 = YCONST - YVAR
                        ENDIF
                ENDIF
        ENDIF
```

2.11 Circles of Given Radius Tangent to a Line and a Circle

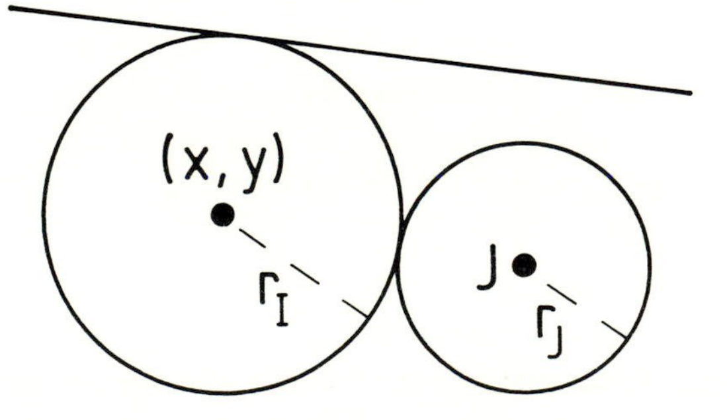

This problem is approached, like that in Section 2.8, by taking the point J as a local origin. The line equation is referred to this origin by transforming the constant term, c:

$$c' = c + ax_J + by_J$$

The signs of the terms in this new equation are important. If the fixed circle crosses the line then the sign of c in the line equation determines whether solutions are found on the same or on the opposite side of the line as the fixed circle centre. The sign of r_J determines whether internal or external tangents are returned:

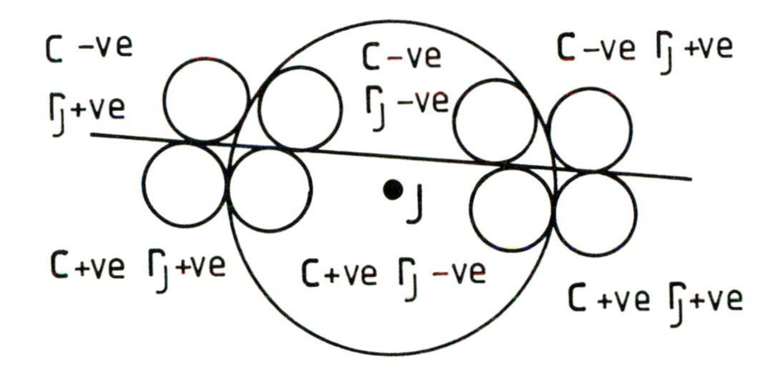

The equation should be multiplied through by -1 to give the sign of c required.

The circle centre coordinates are found from the expressions:

$$x = x_J + \frac{a(c' - r_I) \pm b\sqrt{[(a^2 + b^2)(r_I \pm r_J)^2 - (c' - r_I)^2]}}{(a^2 + b^2)}$$

$$y = y_J + \frac{b(c' - r_I) \mp a\sqrt{[(a^2 + b^2)(r_I \pm r_J)^2 - (c' - r_I)^2]}}{(a^2 + b^2)}$$

which simplify to

$$x = x_J + a(c' - r_I) \pm b\sqrt{[(r_I + r_J)^2 - (c' - r_I)^2]}$$

$$y = y_J + b(c' - r_I) \mp a\sqrt{[(r_I + r_J)^2 - (c' - r_I)^2]}$$

if the line was normalised. The two signs from the root give the two cases on one side of the line.

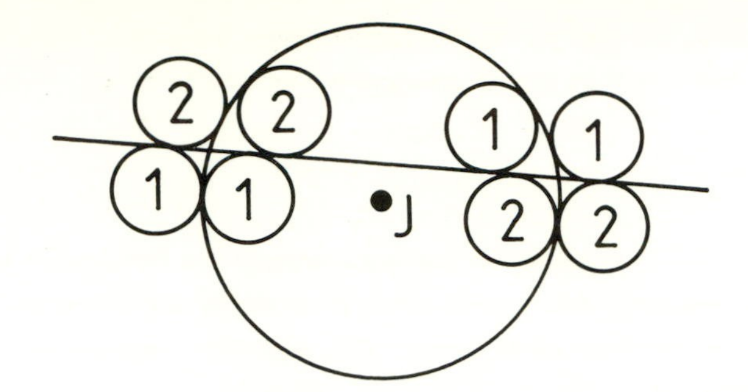

A zero root indicates that only one tangent circle is possible in the specified region. A negative number to be square rooted indicates that J is too far from the line for any tangents to be possible on the side of the line being considered of the type (internal/external) being sought.

Thus, *with the original line normalised* (see Section 1.2), we have code of the form:

```
CDASH = C + A*XJ + B*YJ
IF (CDASH.LT.0.0) THEN                Assumes tangent circles on J side
                                      are required, otherwise use .GE.

        ATEMP = -A
        BTEMP = -B
        CDASH = -CDASH
ELSE
        ATEMP = A
        BTEMP = B
ENDIF
CFAC = CDASH + RI
RFAC = RI + RJ                        Assumes external tangents required
                                      otherwise use RI - RJ

ROOT = RFAC*RFAC - CFAC*CFAC
IF (ROOT.LT.-ACCY) THEN

        ..... There are no solutions in this region

ELSE
        IF (ROOT.LT.ACCY) THEN
                X = XJ + ATEMP*CFAC           Only one circle possible
                Y = YJ + BTEMP*CFAC
        ELSE
                ROOT = SQRT(ROOT)             Two solutions
                XCONST = XJ - ATEMP*CFAC
                YCONST = YJ - BTEMP*CFAC
```

```
                    XVAR = BTEMP*ROOT
                    YVAR = ATEMP*ROOT
                    X1 = XCONST - XVAR
                    Y1 = YCONST + YVAR
                    X2 = XCONST + XVAR
                    Y2 = YCONST - YVAR
            ENDIF
        ENDIF
```

2.12 Circles of Given Radius Tangent to Two Circles

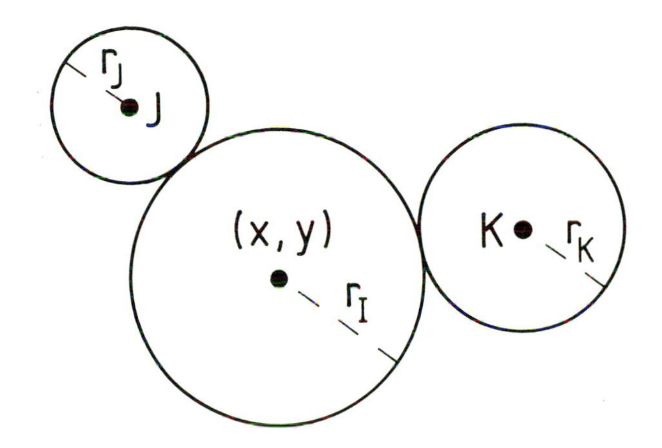

This problem reduces to that given in Section 2.10 by transforming the given radii to make one of the fixed circles into a point with zero radius:

$$r_I' = r_I + r_J$$

$$r_J' = 0$$

$$r_K' = r_K - r_J$$

If this is done the centre of the new circle of radius r_I' will be in the position required for the centre of r_I. If the original radii are signed correctly (positive for external tangency, negative for internal) changes of sign deriving from the transformation will not affect the result.

3

Points, line segments and arcs

3.1 Representation of a Line Segment

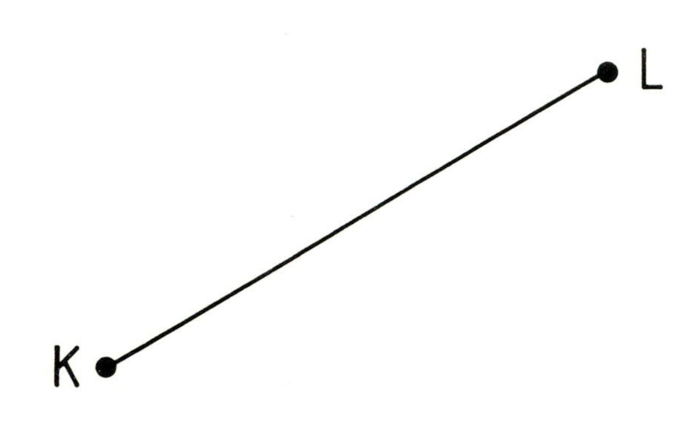

The representation of straight line segments is achieved most simply by storing each segment's endpoints. An alternative is to store the parametric equation of the infinite straight line and the values of the parameter, t, which bound the segment. This is of particular value where a number of segments of the same line are required:

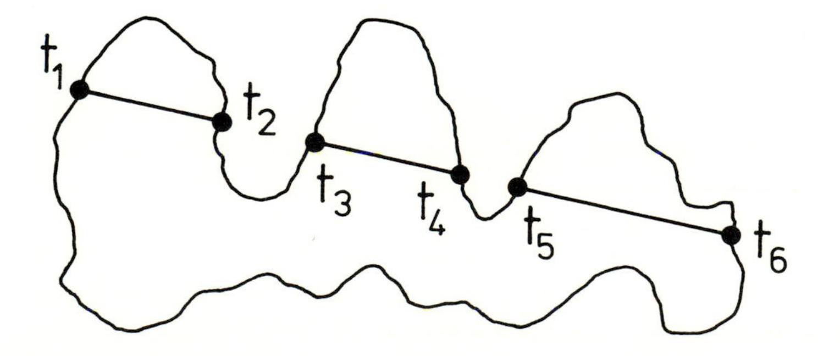

It allows the intersections to be generated in any order and then sorted, segments being then described by consecutive pairs of values.

Both these approaches extend simply to lines in three dimensions. The second is useful for holding the intersections of a ray with the objects in a three-dimensional scene.

3.2 Distance from a Point to a Line Segment

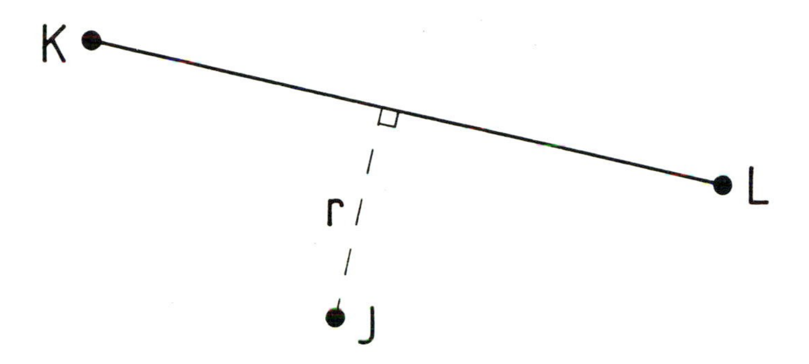

The distance from a point to a line segment is the distance to the line only if the normal from the point to the line strikes the line between the segment endpoints. Otherwise it is the distance from the point to the nearest segment endpoint.

It is convenient to consider the line as the interval t=0 to t=1 of an infinite parametric line:

$$x = x_K + t(x_L - x_K)$$

$$y = y_K + t(y_L - y_K)$$

If we then calculate the value of t where the normal from the point J strikes the line, values between 0 and 1 indicate that the closest point is on the segment, values below 0 indicate that K is the closest point, and values above 1 indicate that L is the closest point. The value of t is found from

$$t = \frac{-[(x_K - x_J)(x_L - x_K) + (y_K - y_J)(y_L - y_K)]}{(x_L - x_K)^2 + (y_L - y_K)^2}$$

and if the nearest point is in the segment, the distance to it is:

$$r = \sqrt{\{[(x_K - x_J) + t(x_L - x_K)]^2 + [(y_K - y_J) + t(y_L - y_K)]^2\}}$$

By truncating the value of t to be between 0 and 1 we can use this formula in all three cases, coded as follows:

```
XKJ = XK - XJ
YKJ = YK - YJ
XLK = XL - XK
YLK = YL - YK
DENOM = XLK*XLK + YLK*YLK
IF(DENOM.LT.ACCY) THEN
        R = SQRT(XKJ*XKJ + YKJ*YKJ)          Segment ends coincide
ELSE
        T = -(XKJ*XLK + YKJ*YLK)/DENOM       Parameter
        T = AMIN1(AMAX1(T,0.0),1.0)          Truncate to ends
        XFAC = XKJ + T*XLK
        YFAC = YKJ + T*YLK
        R = SQRT(XFAC*XFAC + YFAC*YFAC)
ENDIF
```

Dealing separately with the case where K is the nearest point may be slightly more efficient, but with some loss of simplicity. If it is possible to work with squared distances throughout a problem, then the calls to the costly SQRT function can be avoided.

3.3 Intersection of Two Line Segments

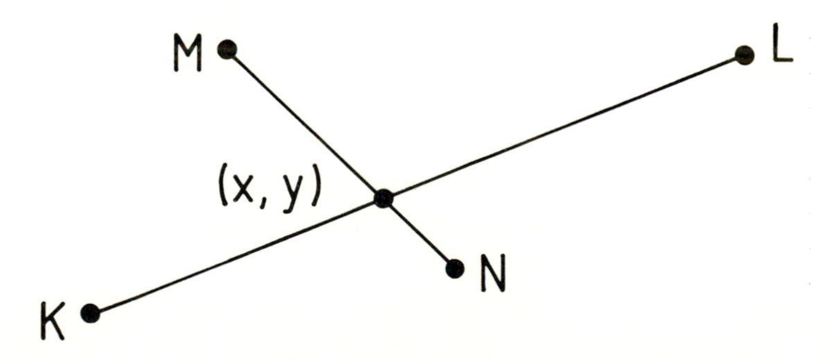

This problem looks deceptively simple. However, the algebra that yields the intersection point directly is inefficient if many of the line segment pairs that are to be compared do not, in fact, intersect. Then it is better to determine whether an intersection is even possible before going on to calculate the coordinates of the intersection. We will first show the direct approach, which in any case leads to the shortest code.

Consider two line segments KL and MN. These will normally be specified by their endpoint coordinates, though if they are specified by the parametric equations of two infinite lines along with two pairs of parameter values the problem is greatly simplified, as such parametric equations have then to be generated from the endpoint coordinates anyway. The endpoints might, for instance, be part of polygons being processed, of which the endpoints would be vertices. One possible way to find the intersection between them is to generate the equations of the corresponding unbounded lines, and find the intersection between them using the method described in Section 1.5. Unfortunately, it is not then particularly easy to decide whether the intersection of the unbounded lines lies within the segments. Instead we determine the parametric equations of the lines in such a way that the line KL has a parameter s running from 0 at K to 1 at L, and MN has a parameter t running from 0 at M to 1 at N. The solution of the set of simultaneous equations then gives the intersection point as:

$$s = \frac{(x_N - x_M)(y_M - y_K) - (y_N - y_M)(x_M - x_K)}{(x_N - x_M)(y_L - y_K) - (y_N - y_M)(x_L - x_K)}$$

$$t = \frac{(x_L - x_K)(y_M - y_K) - (y_L - y_K)(x_M - x_K)}{(x_N - x_M)(y_L - y_K) - (y_N - y_M)(x_L - x_K)}$$

If the values of both s and t are in the range 0 to 1, then the intersection is within both line segments, and the actual coordinates can be found from either of the parameters. For example:

$$x = x_K + (x_L - x_K)s$$

$$y = y_K + (y_L - y_K)s$$

Even if s and t are outside the range 0 to 1, they may still be useful in some circumstances to classify the relationship between the line segments, of which there are eight possible categories:

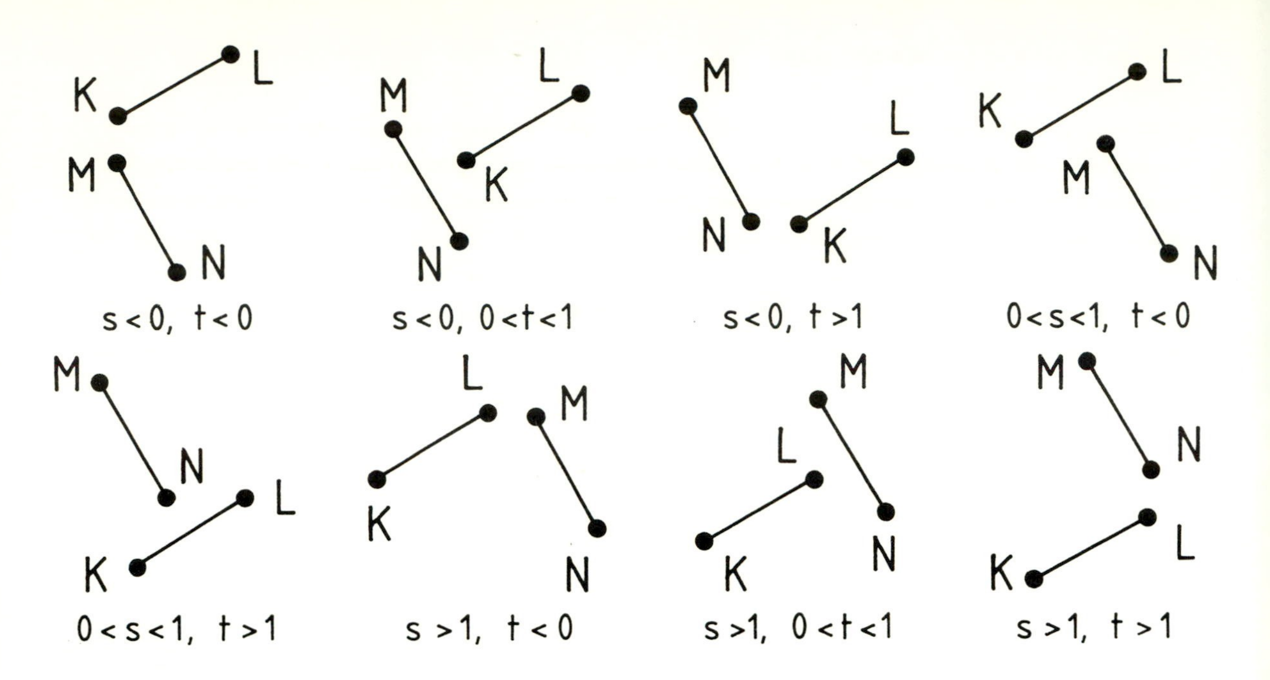

Finding the intersection (if it exists) may be coded:

```
      XLK = XL - XK
      YLK = YL - YK
      XNM = XN - XM
      YNM = YN - YM
      XMK = XM - XK
      YMK = YM - YK

      DET = XNM*YLK - YNM*XLK
      IF(ABS(DET).LT.ACCY) THEN

              ..... The two line segments are parallel

      ELSE
              DETINV = 1.0/DET
              S = (XNM*YMK - YNM*XMK)*DETINV
              T = (XLK*YMK - YLK*XMK)*DETINV
              IF (S.LT.0.0.OR.S.GT.1.0.OR.
     &                T.LT.0.0.OR.T.GT.1.0) THEN

                      ..... Intersection not within line segments

              ELSE
                      X = XK + XLK*S
                      Y = YK + YLK*S
```

```
            ENDIF
        ENDIF
```

If this procedure is being used to find the intersections of many line segments with one single line segment then two of the differences (eg XLK and YLK) need not be recomputed for each comparison.

If an appreciable number of the lines being compared are either horizontal or vertical, then it will be worth branching to much simpler comparisons with such lines. For instance, if the line MN is vertical

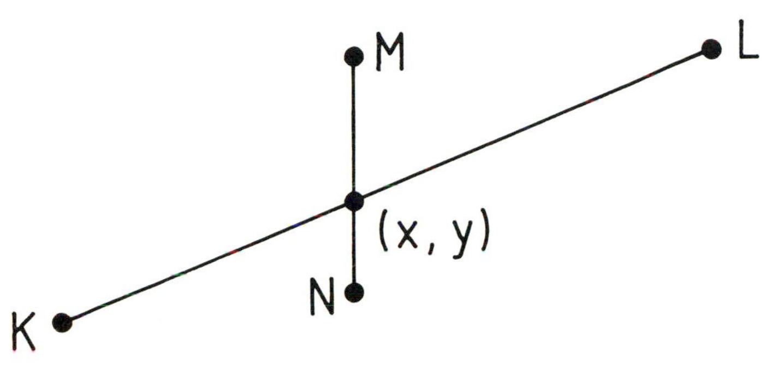

then the intersection point is given by

$$x = x_M$$

$$y = y_K + \frac{(x_M - x_K)(y_L - y_K)}{(x_L - x_K)}$$

which saves seven multiplications and a division over the general method. The value of y can be used without difficulty to determine whether the intersection point is within the line segments. When coding this check for near-zero values of $(x_L - x_K)$; this corresponds to a near vertical line KL.

Unless the reader has a lot of line segments to compare, and is therefore particularly concerned with efficiency, ignore the following; the preceding methods should be quite adequate. In many cases, such as finding the intersections between two polygons, we are concerned to test a single line (one of the sides of the first polygon) against a set of lines (all of the sides of the second polygon). This was mentioned briefly above. Often, when we do this, we expect the majority of the comparisons to yield no intersections between the segments; the line segments may be very distant. In this case the first method does a great deal of unnecessary arithmetic. We will now show how to cull most of these trivial cases by direct comparisons of the endpoints involving only subtractions. This is, in effect, a variation on the idea of *boxing tests*, discussed later in Section 3.7. The effectiveness of tests of this sort depends on the orientation of the line segments in the coordinate system. For the method to be described we can only *guarantee* to reject the possibility of intersection when circles drawn with each line segment as a diameter are disjoint.

Consider the lines KL and MN as above. Let KL be the candidate to be compared with a number of different lines MN. The technique is to classify the endpoints of MN into nine regions with respect to KL. The regions are oriented with KL, not with the coordinate directions, and so there are four non-degenerate (i.e. not vertical or horizontal) cases, depending on the orientation of KL:

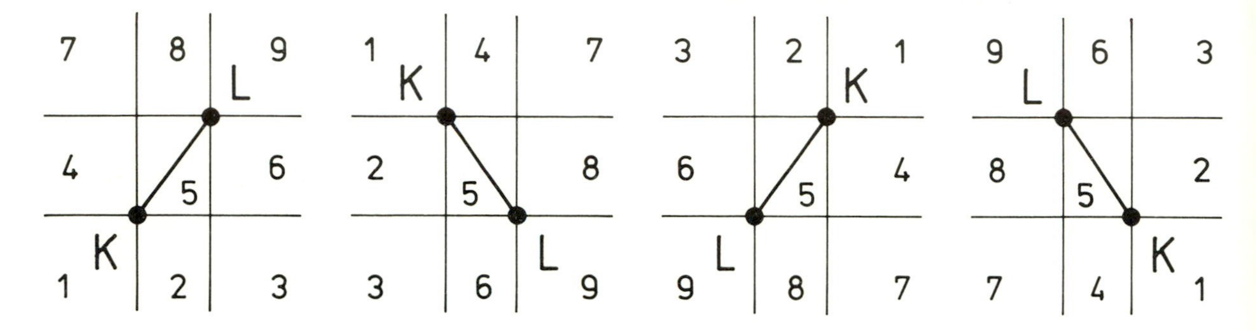

If both point M and point N fall into region 1, then there is clearly no intersection. If they are both in 5, then they must be tested again. In certain cases, such as M in region 5 and N in region 7, then only one further test need be made (on M in this case). In all cases (even if both M and N are in region 5) we have some information about the situation, and need not start the procedure given at the beginning of this section. In fact, all the cases where further information is required can be satisfied by the test to discover on which side of an unbounded line a point lies. The possible outcomes, and the further tests required to resolve inconclusive outcomes, can be tabulated and hence coded as a lookup table:

Region in which point N lies	Region in which point M lies: 1	2	3	4	5	6	7	8	9
1	1	1	1	1	12	3	1	4	7
2	1	1	1	4	8	1	4	2	5
3	1	1	1	4	8	1	7	5	1
4	1	3	3	1	9	2	1	1	6
5	12	10	10	11	12	10	11	11	12
6	4	1	1	2	8	1	5	5	1
7	1	3	7	1	9	6	1	1	1
8	3	2	6	1	9	6	1	1	1
9	7	6	1	5	12	1	1	1	1

The outcomes 1 to 12 correspond to further tests as follows:

1	lines definitely do not intersect
2	lines definitely intersect
3	lines intersect if K is on left of MN
4	lines intersect if K is on right of MN
5	lines intersect if L is on left of MN
6	lines intersect if L is on right of MN
7	lines intersect if K and L are on opposite sides of MN
8	lines intersect if M is on left of KL
9	lines intersect if M is on right of KL
10	lines intersect if N is on left of KL
11	lines intersect if N is on right of KL
12	lines intersect if M and M are on opposite sides of KL

The test to decide which side of a line segment a point is on is derived from the equation for the area of a triangle (Section 4.1). Suppose that it is necessary to find which side of KL the point M lies. This can be found from:

$$r = (x_K - x_M)(y_L - y_M) + (x_L - x_M)(y_K - y_M)$$

If r is negative, then the point is to the left of the line, if r is positive then the point is to the right. Note that r is not the true distance from the point to the line, but is proportional to it. If we have calculated the two values of r for a point to the left and a point to the right of a line the absolute values of r can be used proportionately to divide the distance between the two points, thus giving the coordinates of the point of intersection.

3.4 Representation of an Arc

Arcs are more complicated to represent than line segments. It is possible to store the endpoints and the radius of an arc. The ambiguity in this representation

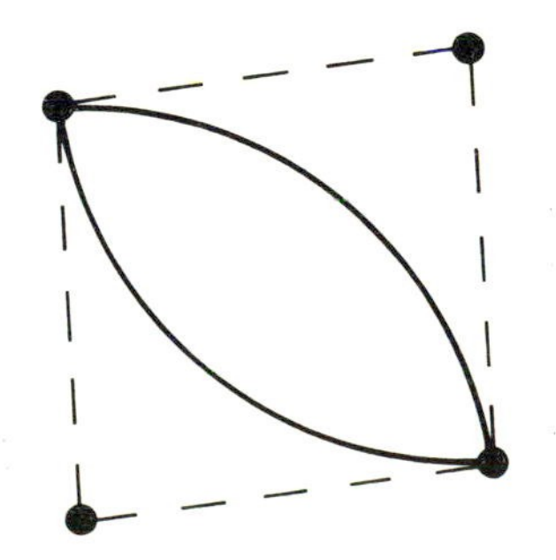

can be avoided by a conventional sign attached to the value of the radius. Clearly the points must not be further than twice the radius apart, as then it would be impossible to draw the arc. Also the arc must never subtend an angle greater than 180°, otherwise ambiguity is reintroduced. This representation can be useful, particularly for the input of contours from, say, some type of digitising device such as a tablet. The problem with it is that it does not facilitate the calculation of the centre, which is essential for most applications (see Section 2 .7). It is possible to add the centre to the data held describing the arc. This corresponds well to some numerically controlled machine tool languages, for instance, but it is bulky, and the consistency of an arc's parameters must be checked; the three points need not represent an arc at all.

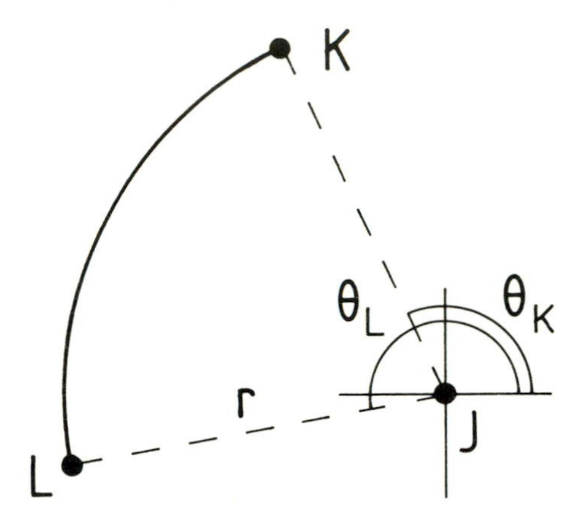

An obvious alternative is to store the arc's centre, radius, and two angles corresponding to the end points. So long as a conventional direction is established (preferably anticlockwise) this allows arcs through any angle with no danger of inconsistency or ambiguity. The drawback is that the sine and cosine functions must be used, at some expense in processor time, to calculate the endpoint coordinates.

$$x_K = x_J + r\cos(\theta_K)$$

$$y_K = y_J + r\sin(\theta_K)$$

As an alternative, the tangent of the half-angle, $\tan(\theta/2)$, may be stored for each end of the arc. This corresponds to the uneven but computationally efficient circle parameterisation mentioned in Section 2.1. To maintain the parameter $\tan(\theta/2)$ in the range 0 to 1, only one quadrant of the circle can be represented. The other quadrants must be indicated by some convention. A compact scheme is to increment the stored value of $\tan(\theta/2)$ with integer values to indicate which quadrant is meant.

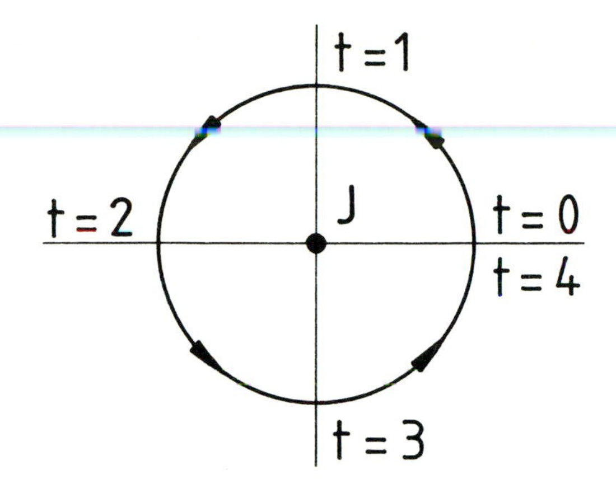

The x and y values are now rather more cheaply calculated by the following code, where TK is the value of tan(θ/2) for the first point, K, added to the integer code for the quadrant:

```
T = AMOD(TK,1.0)                        Remove quadrant information
TSQ = T*T
RFAC = R/(1.0 + TSQ)
DX = (1.0 - TSQ)*RFAC
DY = (T + T)*RFAC
IF(TK.LE.1.0) THEN
        XK = XJ + DX                    First quadrant
        YK = YJ + DY
ELSE IF(TK.LE.2.0) THEN
        XK = XJ - DY                    Second quadrant
        YK = YJ + DX
ELSE IF(TK.LE.3.0) THEN
        XK = XJ - DX                    Third quadrant
        YK = YJ - DY
ELSE
        XK = XJ + DY                    Fourth quadrant
        YK = YJ - DX
ENDIF
```

3.5 Distance from a Point to an Arc

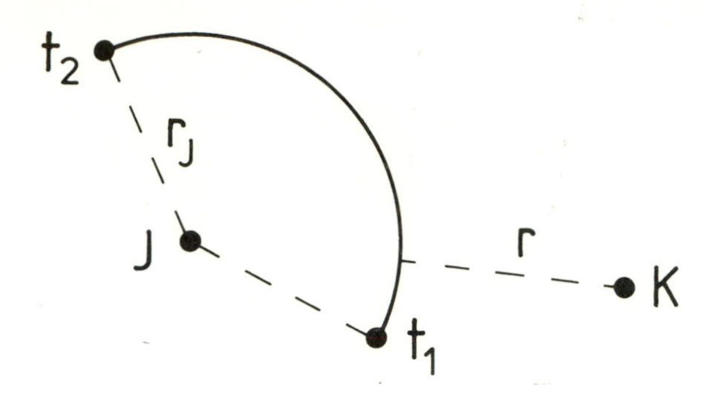

This problem is approached in the same way as the problem of finding the distance to a line segment; the parameter value corresponding to the perpendicular to the whole circle is found and this is compared to the start and finish values of the arc. If the parameterisation is based on tangents of half-angles then trigonometric functions can be avoided by finding the tangent of the angle made by KJ with the x axis and converting this to the half-angle parameterisation using the formula in Section 2.1. Changes in quadrant must also be marked by adding the appropriate integer to the value of t:

```
XKJ = XK - XJ
YKJ = YK - YJ

IF (XKJ.GE.0.0.AND.YKJ.GE.0.0) THEN
        IF (XKJ.GT.ACCY) THEN
                TANT = YKJ/XKJ                          First quadrant
                T = (SQRT(1.0 + TANT*TANT) - 1.0)/TANT
        ELSE
                T = 0.0
        ENDIF

ELSE IF (XKJ.LE.0.0.AND.YKJ.GE.0.0) THEN
        IF (YKJ.GT.ACCY) THEN
                TANT = -XKJ/YKJ                         Second quadrant
                T = 1.0 + (SQRT(1.0 + TANT*TANT) - 1.0)/TANT
        ELSE
                T = 1.0
        ENDIF

ELSE IF (XKJ.LE.0.0.AND.YKJ.LE.0.0) THEN
        IF (XKJ.LT.-ACCY) THEN
                TANT = YKJ/XKJ                          Third quadrant
                T = 2.0 + (SQRT(1.0 + TANT*TANT) - 1.0)/TANT
        ELSE
```

```
                T = 2.0
        ENDIF
ELSE
        IF (YKJ.LT.-ACCY) THEN
                TANT = -XKJ/YKJ                     Fourth quadrant
                T = 3.0 + (SQRT(1.0 + TANT*TANT) - 1.0)/TANT
        ELSE
                T = 3.0
        ENDIF
ENDIF

IF (T2.LT.T1) THEN
        IF (T.GT.T1.OR.T.LT.T2) THEN

                ..... Nearest point is on the arc

        ELSE

                ..... Nearest point is an endpoint

        ENDIF
ELSE
        IF (T.GT.T1.AND.T.LT.T2) THEN

                ..... Nearest point is on the arc

        ELSE

                ..... Nearest point is an endpoint

        ENDIF
ENDIF
```

If the nearest point is on the arc then the distance from the point to the arc is simply:

$$r = \sqrt{[(x_K - x_J)^2 + (y_K - y_J)^2]} - r_J$$

A negative distance indicates that J is within the circle.

Conversely, if the nearest point to K is an endpoint, then the distance to both endpoints must be calculated using the code in Section 3.4 to find the x and y coordinates corresponding to T1 and T2. The minimum of these two distances is taken as the answer.

3.6 Intersections of a Line Segment and an Arc

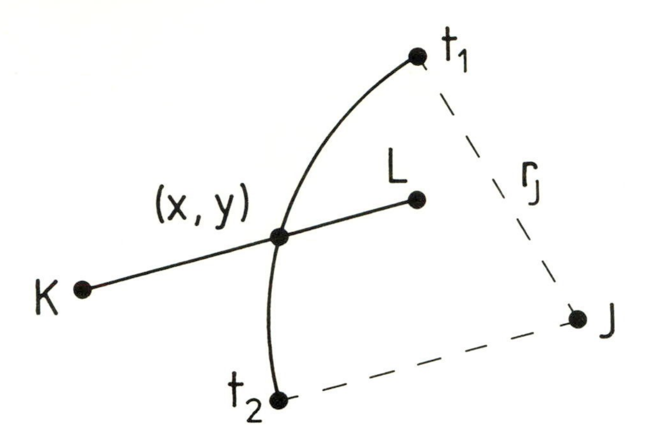

The direct solution of this problem starts with the determination of the intersections of the infinite line of which the segment is a part with the whole circle of which the arc is a part. This operation is covered in Section 2.2. If the infinite line and the whole circle do not intersect, rejection of the arc/segment pair is quick. If, however, the line segment and arc are both short compared to the circle radius, many cases where the segment and the arc are quite distant from each other will remain unrejected until quite late in the computation.

We therefore use a pre-test as we did for segment-segment intersections in Section 3.3; this time an orthodox *boxing test*. Notional rectangles with sides parallel to the axes are constructed and these are compared before comparing the segment and arc that they contain. If they do not intersect the segments may be rejected immediately.

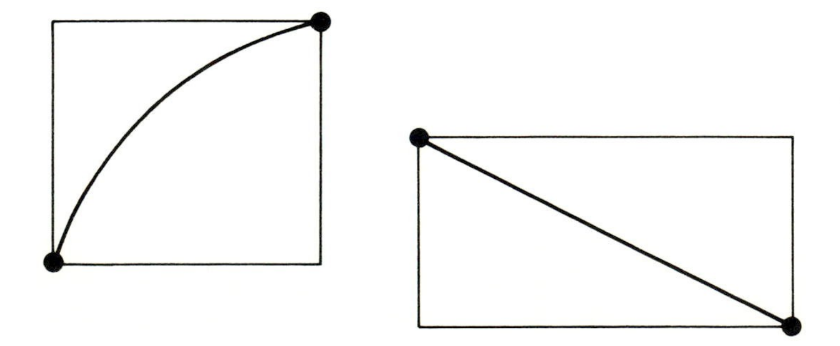

If the boxes do intersect, a more detailed test is necessary to determine whether or not the segment and the arc themselves intersect.

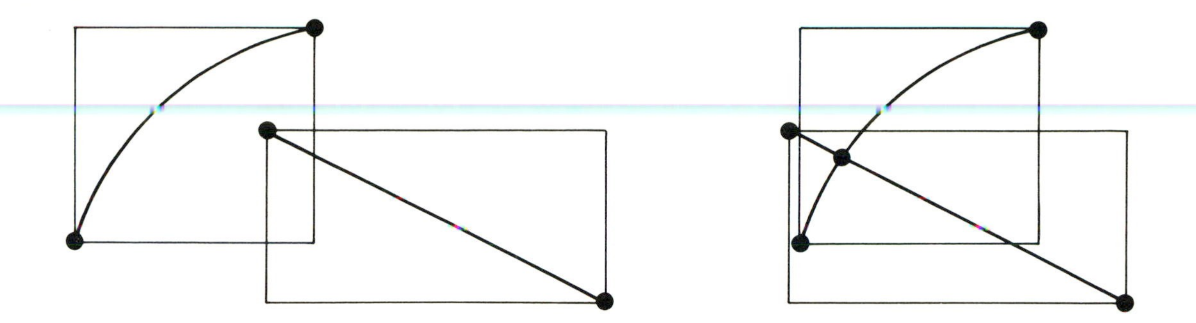

The idea is that most rejections may be made early, especially when segments are small and well separated, which is, of course, the worst case for direct comparison.

For line segments, the corners of the box corresponded to the segment ends. The same is true for arcs if the arc is in a single quadrant. If the arc occupies several quadrants the coordinates of the extreme x and y values of the circle become the x and y coordinates of one or more edges of the box. For example:

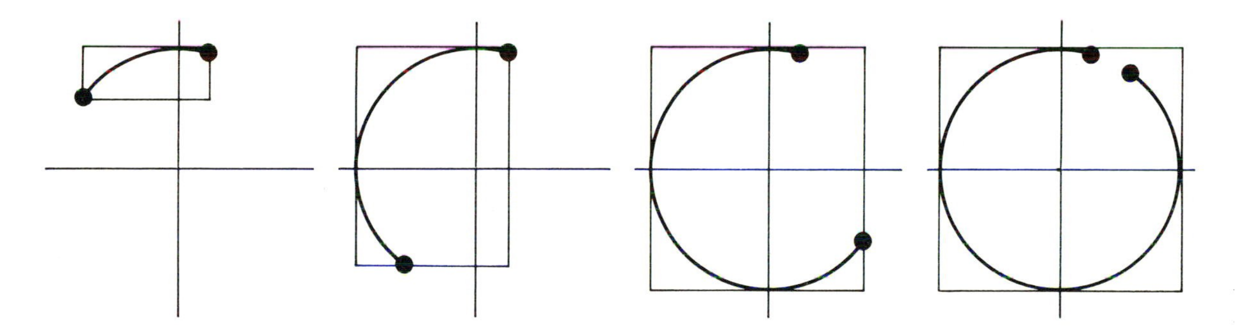

The following table shows which values must be used to extend the box for given start and finish quadrants of the arc.

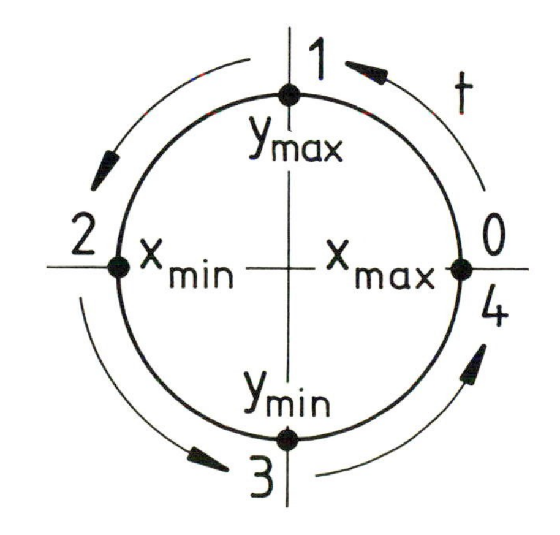

Arc ends \ Arc begins	t = 0–1	t = 1–2	t = 2–3	t = 3–4
t = 0–1	XMIN YMIN XMAX YMAX	XMIN YMIN XMAX	YMIN XMAX	XMAX
t = 1–2	YMAX	XMIN YMIN XMAX YMAX	YMIN XMAX YMAX	XMAX YMAX
t = 2–3	XMIN YMAX	XMIN	XMIN YMIN XMAX YMAX	XMIN XMAX YMAX
t = 3–4	XMIN YMIN YMAX	XMIN YMIN	YMIN	XMIN YMIN XMAX YMAX

If arcs are to be subjected to repeated boxing tests, then storing their endpoints will save computing time.

If a boxing test fails to eliminate a comparison the next step is to calculate the intersections of the line and the whole circle. Even after a boxing test no intersections may exist. Then it is necessary to parameterise the line segment as described in Section 1.6, and then to work out the intersections (if any) of the resulting parametric line with the *whole* circle, as shown in Section 2.2. Line parameters in the range 0 to 1 at the intersections correspond to intersections inside the segment. The corresponding x and y values are used to calculate the tangent of the angle made by the candidate intersection at the circle centre. This tangent can be converted to a half-angle tangent using the formula:

$$\tan\frac{\theta}{2} = \frac{\sqrt{(1 + \tan^2\theta)} - 1}{\tan\theta}$$

This value can be compared with the values of the arc parameters at the arc ends to see if the intersection lies within the arc or not. The coding of this is identical to the classification of the perpendicular intersection in section 3.5.

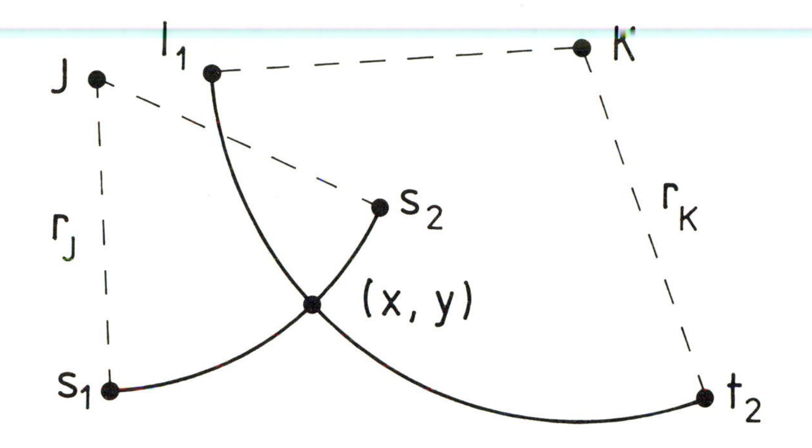

The various operations required for the solution of this problem have already been covered in earlier sections of this and other chapters. If efficiency is at a premium boxing tests may be performed on the two arcs in the same way as for the arc in Section 3.6.

Following any boxing tests, the coordinates of the intersections between the whole circles are found, as described in Section 2.3. If no intersections exist between the whole circles then intersection between the arcs is ruled out. The two sets of intersection coordinates (or only one, if the circles are found to be tangential to each other) are referred to each arc centre in turn to convert them to the half-angle arc parameterisation, following the procedure used in Section 3.5. This section also shows how the values of the parameter thus obtained may be used to determine whether or not the intersections lie on the arcs.

Intersections lying on both arcs are valid: there may be one or two true intersections, or one tangent point.

4 Areas

4.1 Area of a Triangle

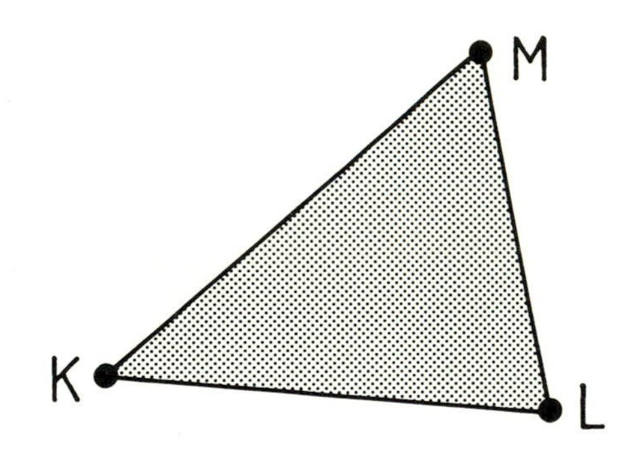

The triangle is particularly useful in computer geometry because it is simply and unambiguously described by the coordinates of its three vertices. It also has the merit of always being planar, even when the vertices are points in space. It is therefore common to break down two-dimensional shapes and three-dimensional surfaces into triangles, so that programs working with these structures have only one element to handle.

The area of a triangle is found by evaluating the determinant (see Section 6.3)

$$\frac{1}{2}\begin{vmatrix} (x_L - x_K) & (x_M - x_K) \\ (y_L - y_K) & (y_M - y_K) \end{vmatrix}$$

which may be coded:

```
XLK = XL - XK
XMK = XM - XK
YLK = YL - YK
YMK = YM - YK

AREA = 0.5*(XLK*YMK - XMK*YLK)
AREA = ABS(AREA)
```

If the absolute value of the area is not taken, its sign indicates the relationship between point K (in general, the subtracted point) and the line segment defined by the other two points. If the determinant is positive, K, L, and M are in anti-clockwise order; if it is negative they are clockwise. Thus without the call to the ABS function and without the factor of 0.5 this code gives a quick way to decide on which side of a line segment a given point lies.

An alternative area formula is in terms of the side lengths

$$\sqrt{[s(s - r_{KL})(s - r_{LM})(s - r_{MK})]}$$

where r_{KL} and so on are the side lengths, and s is half the the length of the triangle perimeter.

To find the area of a triangle in three dimensions use the formula given at the end of Section 8.2.

4.2 Centre of Gravity of a Triangle

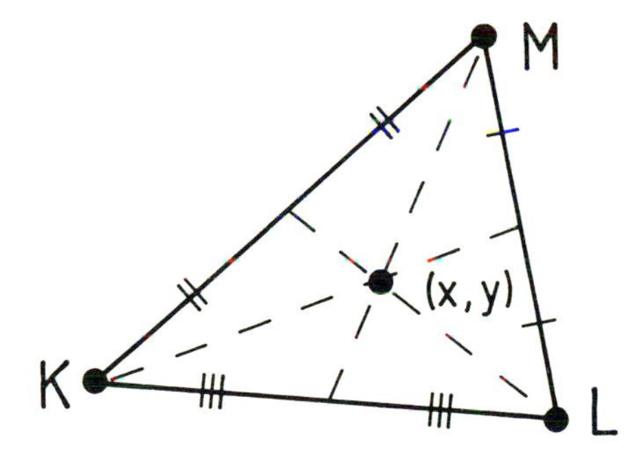

The centroid of a triangle lies at the meeting point of lines drawn from each vertex to the mid-point of the opposite side. It is, however, simply calculated as the average of the vertex coordinates:

$$x_{CG} = \frac{x_K + x_L + x_M}{3}$$

$$y_{CG} = \frac{y_K + y_L + y_M}{3}$$

4.3 Incentre of a Triangle

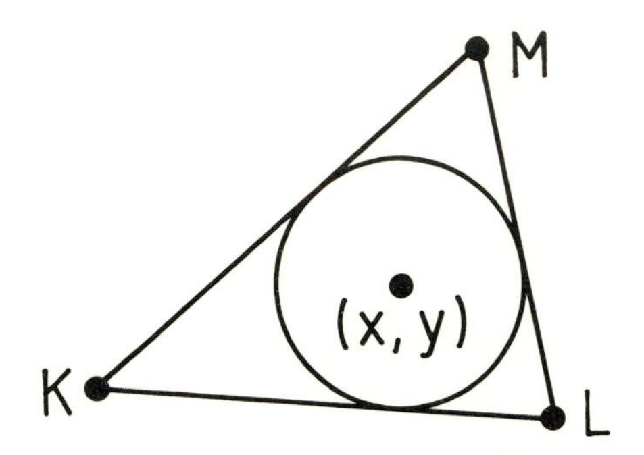

The inscribed circle is the largest that can be drawn inside a triangle. It is tangent to all three sides, and the lines from its centre to the vertices bisect the angle made by the triangle's sides at each vertex.

The position of the incentre is given by

$$x_{IN} = \frac{r_{LM}x_K + r_{MK}x_L + r_{KL}x_M}{t}$$

and

$$y_{IN} = \frac{r_{LM}y_K + r_{MK}y_L + r_{KL}y_M}{t}$$

where r_{KL} and so on are the lengths of the sides of the triangle and t is its perimeter. The radius of the inscribed circle is

$$R_{IN} = \sqrt{\left(\frac{(s - r_{LM})(s - r_{MK})(s - r_{KL})}{s}\right)}$$

where s = t/2.

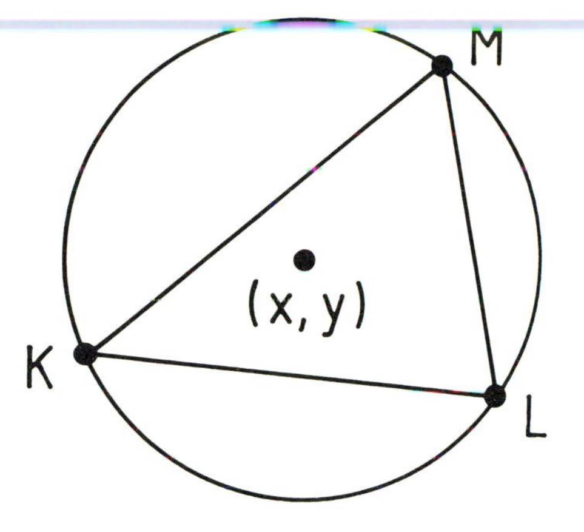

The circumcentre of a triangle is the centre of the circle that passes through the vertices of the triangle. It is found by treating one vertex temporarily as the origin, when it becomes

$$x_{CC} = \frac{\begin{vmatrix} (x_{LK}^2 + y_{LK}^2) & y_{LK} \\ (x_{MK}^2 + y_{MK}^2) & y_{MK} \end{vmatrix}}{2 \begin{vmatrix} x_{LK} & y_{LK} \\ x_{MK} & y_{MK} \end{vmatrix}}$$

$$y_{CC} = \frac{\begin{vmatrix} x_{LK} & (x_{LK}^2 + y_{LK}^2) \\ x_{MK} & (x_{MK}^2 + y_{MK}^2) \end{vmatrix}}{2 \begin{vmatrix} x_{LK} & y_{LK} \\ x_{MK} & y_{MK} \end{vmatrix}}$$

(see Section 6.3 for a brief explanation of determinants) where

$x_{LK} = x_L - x_K$, $\quad x_{MK} = x_M - x_K$

and $\quad y_{LK} = y_L - y_K$, $\quad y_{MK} = y_M - y_K$

The centre given by (x_{CC}, y_{CC}) is relative to the position of K, so the coordinates of that point have to be added to get the absolute position of the centre. The relative position can be used to find the squared radius before this addition is performed:

$$r_{CC}^2 = x_{CC}^2 + y_{CC}^2$$

This is coded:

```
XLK = XL - XK
YLK = YL - YK
XMK = XM - XK
YMK = YM - YK
DET = XLK*YMK - XMK*YLK
IF (ABS(DET).LT.ACCY) THEN

          ..... At least two of the points are coincident

ELSE
          DETINV = 0.5/DET
          RLKSQ = XLK*XLK + YLK*YLK
          RMKSQ = XMK*XMK + YMK*YMK
          XCC = DETINV*(RLKSQ*YMK - RMKSQ*YLK)
          YCC = DETINV*(XLK*RMKSQ - XMK*RLKSQ)
          RCCSQ = XCC*XCC + YCC*YCC
          X = XCC + XK
          Y = YCC + YK
ENDIF
```

4.5 Representation of a Polygon

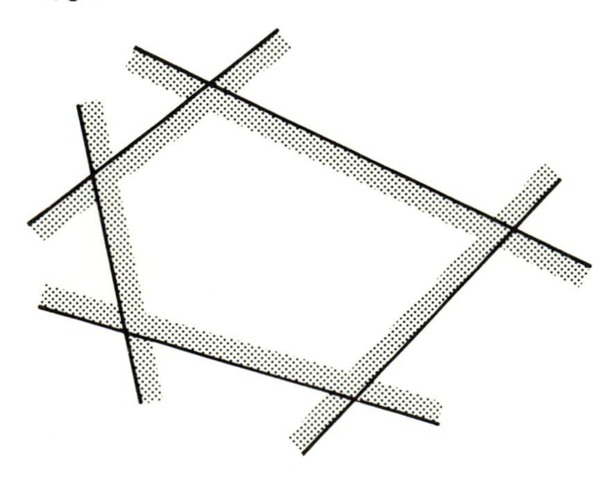

The simplest way to define a *convex* polygon (in other words a polygon where all the internal angles are less than 180°) is to consider each of the sides as a linear half-plane (see Section 1.2), with its true side pointing inwards. The polygon is the region on the true side of all the half-planes. As a consequence of this decision, it is easy to determine whether or not a given point is inside the polygon. The point coordinates are substituted into each half-plane equation in turn. If all these substitutions give a negative result (by the convention in Section 1.2), the point is inside the polygon. If any result is positive, it is outside.

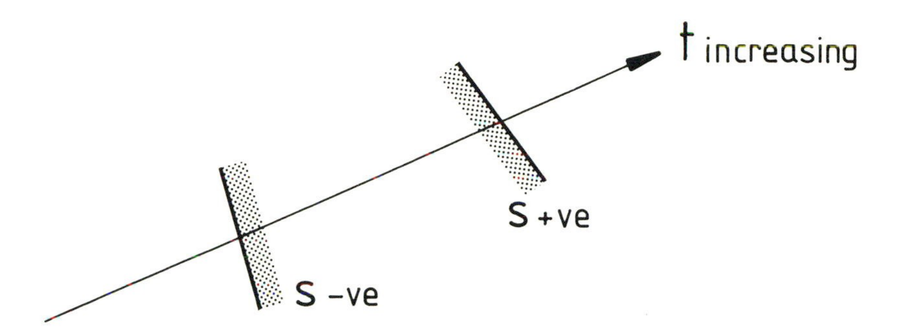

The half-planes may be converted to an ordered list of vertices as follows. Each half-plane is converted to parametric form (see Section 1.2), and its intersections with the other half-planes is found (see Section 1.5). For each intersection the parameter value on the candidate line is generated. By taking the scalar product (Section 6.1) of the half-plane normal and the parametric line slope coefficients

$$s = fa + gb$$

the intersections may be classified as including the *half-line* with parameter increasing or decreasing (we assume an outward pointing normal on the half-planes).

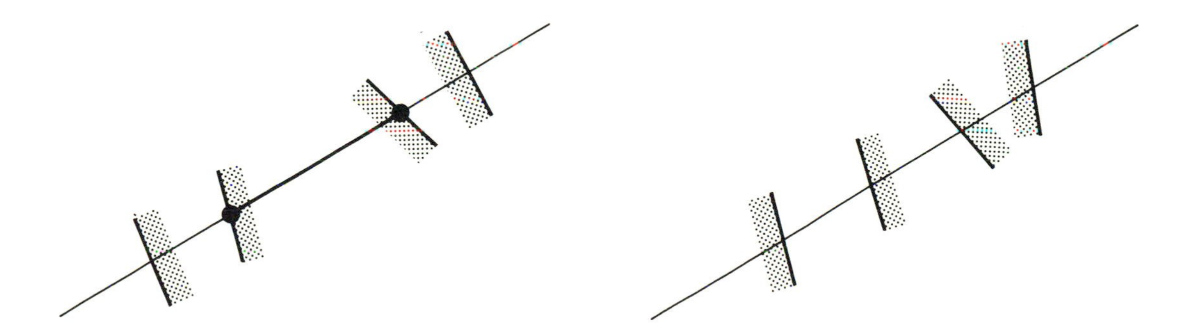

If the lowest intersection which includes the half-line with parameter decreasing is parametrically greater than the highest intersection which includes the half line with parameter decreasing, then the two intersections describe a side of the polygon. Otherwise the half-plane does not contribute to the polygon at all.

As each segment is generated, the identity of the half-plane which created each end is noted. When all the segments have been generated this information is used to trace round the segments in order, and hence to generate the list of vertex coordinates.

An ordered list of vertices is a more common and slightly more compact (two-thirds of the space) way of storing polygons, but it does not ensure that the polygon is convex. However, it is useful in many applications, the simplest being to draw the polygon.

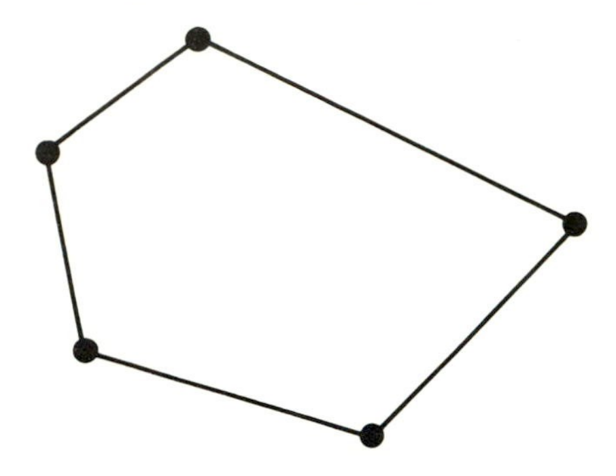

The problem of determining whether a point is inside the polygon is now more complex. It is necessary to construct a line or *ray* from the point to be tested to infinity and to determine whether this crosses the sides of the polygon. If it does not cross the sides, or crosses them twice, the point is outside. If it crosses them once, it is inside.

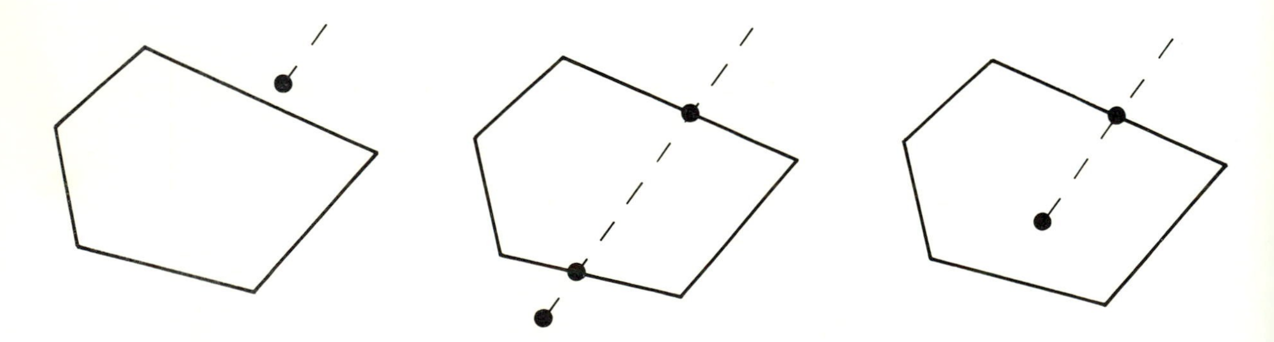

This test suffers accuracy problems when the ray passes near a vertex. This occurrence must be detected and a new ray chosen.

The half-plane representation of a convex polygon may be written as the set-theoretic *intersections* (∩) of the regions defined by the half-planes: each region being considered to be made up of an infinite set of points. By using the additional set-theoretic operators union (∪) and difference (⊥) any polygon can be created by adding and subtracting convex polygons.

The ordered list of vertices is easily extendible to non-convex polygons; it has already been said that it does not ensure convexity in any case. The ray test may be used to identify the inside and outside regions of such a polygon, odd numbers of intersections with edges indicating that the candidate point is inside the polygon, even numbers (including zero) indicating outside.

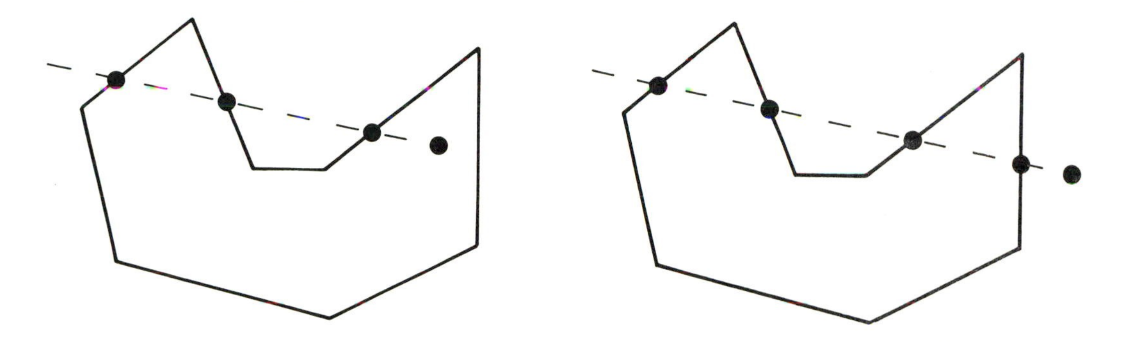

If the ordered list representation is used, care must be taken that the polygon sides do not cross each other. This condition invalidates the algorithms given in Sections 4.6 and 4.7.

4.6 Area of a Polygon

The area of any polygon represented as a vertex list may be calculated by summing the areas of the trapezia under each side, down to the axis. The direction of the sides must be taken into account, so that sides on the bottom of the polygon are subtracted from the total.

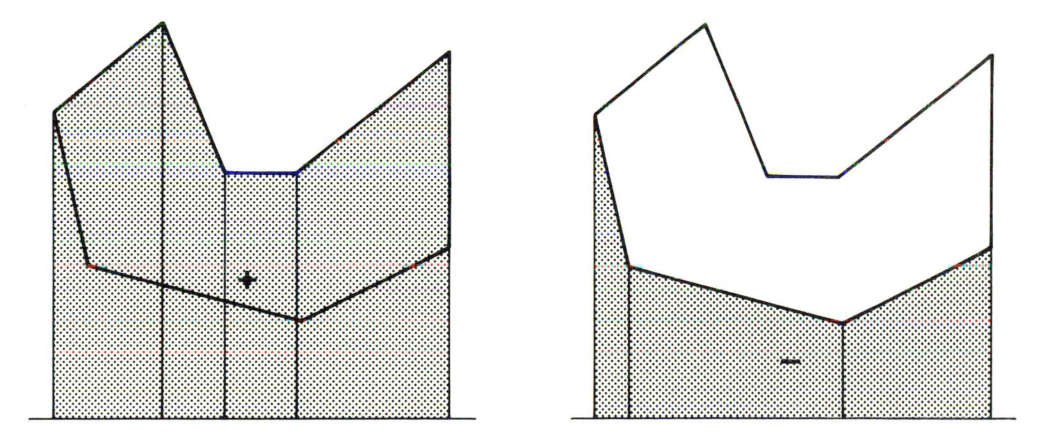

Suppose that the real arrays XVERT and YVERT hold the NVERT coordinates of a polygon's vertices, stored in anti-clockwise order. The area of the polygon may be calculated using the following code:

```
AREA = 0.0
XOLD = XVERT(NVERT)
YOLD = YVERT(NVERT)
```

```
      DO 10 N = 1, NVERT
              X = XVERT(N)
              Y = YVERT(N)
              AREA = AREA + (XOLD - X)*(YOLD + Y)
              XOLD = X
              YOLD = Y
   10 CONTINUE

      AREA = 0.5*AREA
```

If it is not known whether the polygon was stored in anticlockwise or clockwise order then the absolute value of the area should be taken.

There is one major problem with this approach. If the polygon is a long way from the x axis then the area of the trapezia will be much larger than the area of the polygon and accuracy will be lost. Temporarily making one vertex the y origin will avoid this problem:

```
      AREA = 0.0
      XOLD = XVERT(NVERT)
      YORIG = YVERT(NVERT)
      YOLD = 0.0

      DO 10 N = 1,NVERT
              X = XVERT(N)
              Y = YVERT(N) - YORIG
              AREA = AREA + (XOLD - X)*(YOLD + Y)
              XOLD = X
              YOLD = Y
   10 CONTINUE

      AREA = 0.5*AREA
```

4.7 Centre of Gravity of a Polygon

An alternative approach to computing the area of a polygon is to take a point and construct triangles by joining the point to all the vertices of the polygon. The point need not lie in the polygon. The area is then found by summing the areas of the triangles, again signed according to edge direction. If a vertex of the polygon is chosen as the point the number of triangles that have to be dealt with is two less than in the general case.

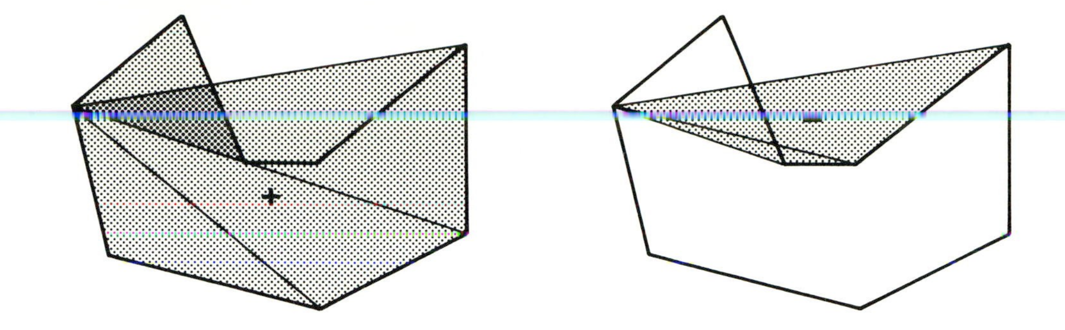

It so happens that this approach is computationally slightly less efficient than the trapezoidal method for area calculations, but more efficient when the centre of gravity of the polygon is required.

The formulae for the area and centre of gravity of a triangle have already been described (Sections 4.1 and 4.2). Using the same arrays, XVERT and YVERT, that were used to find areas in Section 4.6, and taking vertex NVERT as the common point, the centre of gravity of the polygon may be found as follows. In this case it is essential to know the cyclic direction of the polygon, as the centre of gravity may genuinely have negative coordinates. We again assume that the polygon vertices are stored anti-clockwise.

```
      XCG = 0.0
      YCG = 0.0
      ARESUM = 0.0
      XCOM = XVERT(NVERT)
      YCOM = YVERT(NVERT)
      XOLD = XVERT(1)
      YOLD = YVERT(1)

      NVT1 = NVERT - 1
      DO 10 N = 2, NVT1
              X = XVERT(N)
              Y = YVERT(N)
              ARETRI = (XCOM - X)*(YOLD - YCOM) +
     &                 (XOLD - XCOM)*(Y - YCOM)

              XCG = XCG + ARETRI*(X + XOLD)
              YCG = YCG + ARETRI*(Y + YOLD)
              ARESUM = ARESUM + ARETRI
              XOLD = X
              YOLD = Y
   10 CONTINUE
      IF(ARESUM.LT.ACCY) THEN
```

..... *The polygon covers no area or is defined clockwise*

```
ELSE
          AREINV = 1.0/ARESUM
          XCG = (XCG*AREINV + XCOM)*0.333333
          YCG = (YCG*AREINV + YCOM)*0.333333
ENDIF
```

Note that XCOM and YCOM are not added into the centre of gravity for every triangle and then divided by the area, but that this operation is performed only once at the end. ARESUM is *double* the true area of the polygon, so if both the centre of gravity and the area of a polygon are required a separate calculation for the latter is unnecessary.

4.8 Centre of Gravity of a Sector and a Segment

It is well known that the circumference of a circle is $2\pi r$, and that its area is πr^2. More interesting are the sector and segment:

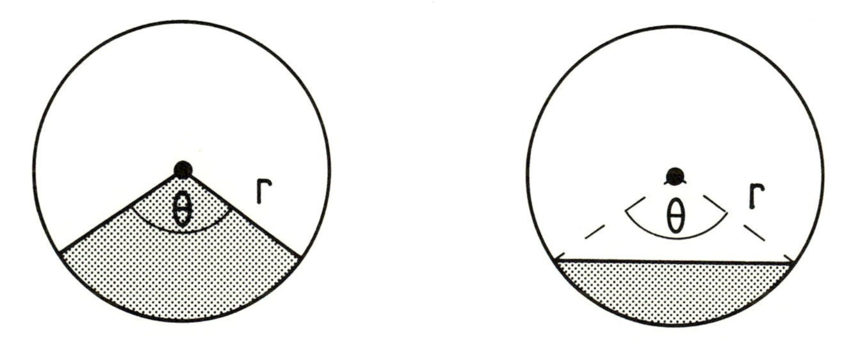

The area of a segment is simply given as a fraction of the area of the circle:

$$\frac{\theta r^2}{2}$$

The area of a sector is obtained by subtracting a triangular piece from this:

$$\frac{r^2(\theta - \sin\theta)}{2}$$

The centres of gravity of both figures lie on the bisector of the central angle, by symmetry. The distance from the circle centre to the centroid is given by

$$\frac{4r\sin(\theta/2)}{3\theta} \quad \text{for the sector}$$

and

$$\frac{4r\sin^3(\theta/2)}{3(\theta - \sin\theta)} \quad \text{for the segment.}$$

Both of these formulae should be used with caution for small values of θ.

5
Curves other than circles

5.1 General Implicit Quadratic Equations

The general equation of a quadratic (note the slight change from our usual notation to accommodate the number of coefficients, and also the factors of 2, which make certain calculations easier) is:

$$ax^2 + 2bxy + cy^2 + 2dx + 2ey + f = 0$$

Such general quadratics are called *conic sections* as they can represent all the shapes that it is possible to get by cutting a cone with a plane. The three shapes that can be obtained in this way are the ellipse (of which the circle is a special case), the parabola and the hyperbola. If we calculate three values from the coefficients:

$$\Delta = a(cf - e^2) + b(bf - de) + d(be - dc)$$

$$\delta = ac - b^2$$

$$S = a + c$$

(We have changed our usual notation here to comply with the usual mathematical conventions for these values.) Then, no matter how the quadratic is moved about in the plane using translation (sliding) and rotation, these three values will stay the same as long as the shape of the quadratic stays the same.

Categorising a given quadratic into one of the three possible forms it can take (ellipse, parabola, or hyperbola) can be done using the three invariants whose calculation was described above. The categorisation is sumarised as follows:

If Δ is 0 then the quadratic is degenerate and represents two straight lines (which may not always exist) otherwise:

$\delta < 0$ Quadratic is a hyperbola

$\delta = 0$ Quadratic is a parabola

$\delta > 0$ Quadratic is an ellipse

In the latter case the ellipse only exists if ΔS is negative. Note that, when coding this, it is unlikely that a calculation of Δ will yield exactly zero for a degenerate quadratic, and similarly δ will be small, but not exactly zero, for a parabola. This imprecision is the result of rounding error, and is unavoidable on a machine that uses floating point arithmetic. General quadratics are dealt with in more detail in Bronshtein and Semendyayev.

The general quadratic can be useful for fitting smooth curves through patterns of given points and tangential to given lines. This is covered in the section on Liming multipliers in Section 5.2.

5.2 Interpolation Using General Implicit Quadratics

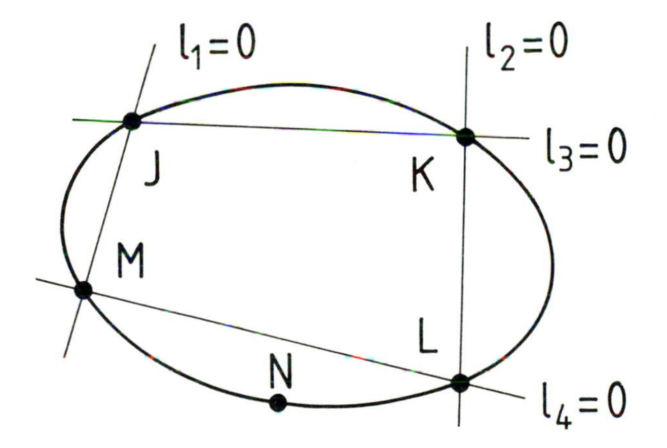

Suppose that we have two pairs of straight lines l_1, l_2, and l_3, l_4 where

$$l_i = a_i x + b_i y + c_i, \quad i = 1, 2, 3 \text{ and } 4$$

and we multiply these line equations together with a factor λ, which is known as a *Liming multiplier*:

$$(1 - \lambda) l_1 l_2 + \lambda l_3 l_4 = 0$$

We will have generated a family of implicit quadratic equations, each different value that we choose for λ giving a different quadratic. All the quadratics will have the property that they will pass through the intersection points of the pairs of lines, J,K,L, and M. To specify one quadratic we need a value for λ. This can be found by specifying a fifth point, N, through which the quadratic must pass and then substituting the value of (x_N, y_N) into the quadratic to find λ. (To find l_1 .. l_4 given J,K,L and M use the method given in Section 1.6.)

If we reduce the number of lines to three by making two of them equal ($l_3 = l_4$, say) then we get quadratics tangential to l_1 and l_2 at the points where l_3 cuts those two lines.

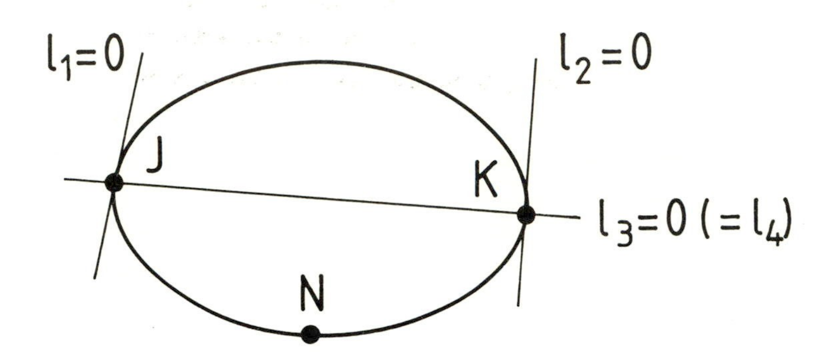

and the quadratics become:

$$(1 - \lambda)l_1l_2 + \lambda l_3^2 = 0$$

Again we can tie down the value of λ by specifying another point, N, through which the quadratic is to pass and substituting its x and y coordinates back into the quadratic.

This technique is very useful for constructing piecewise quadratics that join smoothly (in other words which have common tangents where they join) at given points or with given straight lines. The technique is described more fully in Chapter 1 of Faux and Pratt.

5.3 Parametric Polynomials

Implicit equations which are higher order than quadratic (ie have terms in x^3, x^2y etc.) are not generally useful because of the problems encountered in solving them to obtain a value of y for a given x coordinate, or vice versa. Extending the parametric line to higher orders by adding terms in t^2, t^3 and so on does not give rise to this problem as values of x and y are easily calculated from a value of t. These *parametric polynomials* may therefore be used to generate more flexible curves than is possible with the implicit quadratic.

The simplest non-linear parametric curve is the quadratic:

$$x = a_1 + b_1 t + c_1 t^2$$

$$y = a_2 + b_2 t + c_2 t^2$$

(Note the change in notation from that used in previous parametric equations to accommodate the additional terms.) The next form is the parametric cubic

$$x = a_1 + b_1 t + c_1 t^2 + d_1 t^3$$

$$y = a_2 + b_2 t + c_2 t^2 + d_2 t^3$$

and so on, with higher order equations being formed by adding more terms. Three-dimensional curving lines in space may be formed by adding a third equation in z.

5.4 Interpolation Using Parametric Polynomials

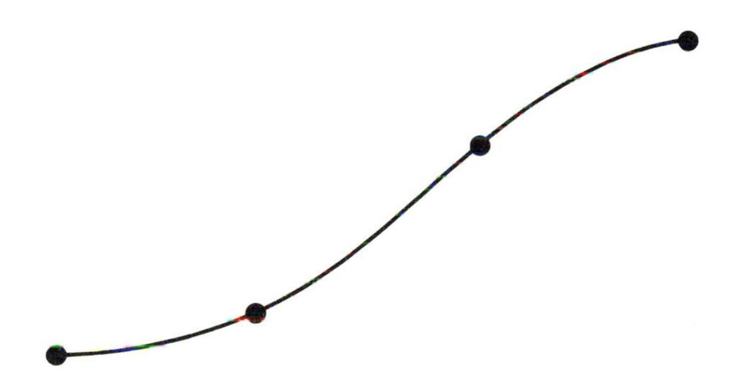

Parametric polynomials are often used to interpolate a curve through a set of data points. To do this it is first necessary to choose the value of t which will correspond to each given point, thus determining the order in which the curve passes through the points. The chosen values of t and the corresponding x and y values for the points are substituted into the parametric equation at each point. This gives two sets of linear simultaneous equations in the coefficients of the parametric polynomials. If the order of the curve (the highest power of t used) is one less than the number of points (3 points for a quadratic, 4 for a cubic etc.), then the simultaneous equations can be solved. The curve is thus defined, and it may then be drawn or used in other calculations.

Programs for the solution of many simultaneous equations are beyond the scope of this book, but the reader employing a large mainframe computer may well find that he has a library package (such as the NAG library) available for the purpose. Users of small computers may not be in such a fortunate position, though a subset of the NAG library is available that runs under the CP/M microcomputer operating system. Alternatively, readers should consult Wilkinson and Reinsch.

Interpolation through points is often called *Lagrangian Interpolation*. *Hermite Interpolation*, on the other hand, is concerned with fulfilling slope constraints as well.

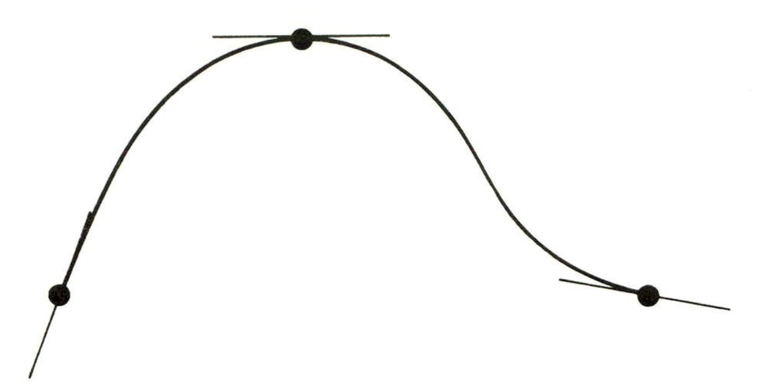

To achieve this, the equations are differentiated. For example a cubic

$$x = a_1 + b_1 t + c_1 t^2 + d_1 t^3$$

$$y = a_2 + b_2 t + c_2 t^2 + d_2 t^3$$

has differentials:

$$\frac{dx}{dt} = b_1 + 2c_1 t + 3d_1 t^2$$

$$\frac{dy}{dt} = b_2 + 2c_2 t + 3d_2 t^2$$

The values of dx/dt and dy/dt are generated from the specified slopes at each given point for the values of t that it has been decided to use at that point. These values are substituted into the differential equations to give yet more simultaneous linear equations in the polynomial coefficients, and the entire set, from original and differentiated polynomials, is solved together to give values for the coefficients.

It is not, of course, necessary to have slope constraints at every point; position and slope constraints can be mixed as required.

Interpolation does not always yield the curve that the reader might intuitively expect. Here are three points to watch:

1) If the initial points are approximately evenly spread along the course of the desired line,

then the points can be parameterised at even intervals of t. If the points are uneven, however, this should be reflected in the parameterisation. Making the intervals in the parameter proportional to the distance between points is a solution commonly adopted:

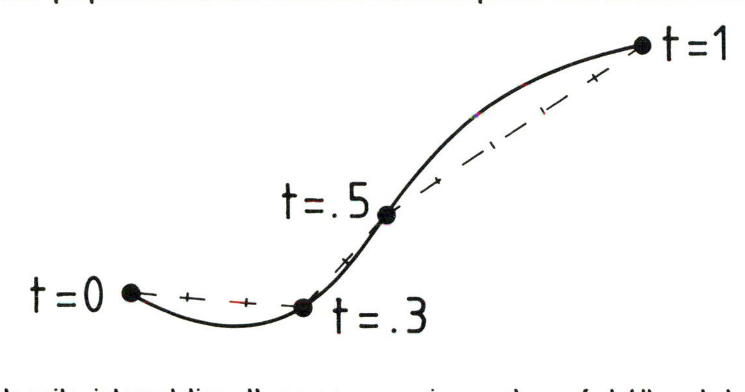

2) In Hermite interpolation, there are no unique values of dx/dt and dy/dt for a required dy/dx (slope), only the ratio of (dy/dt)/(dx/dt) must correspond. Increasing the actual values of dx/dt and dy/dt specified will lead to a flatter curve at the point being considered, but may perhaps produce unwanted effects elsewhere.

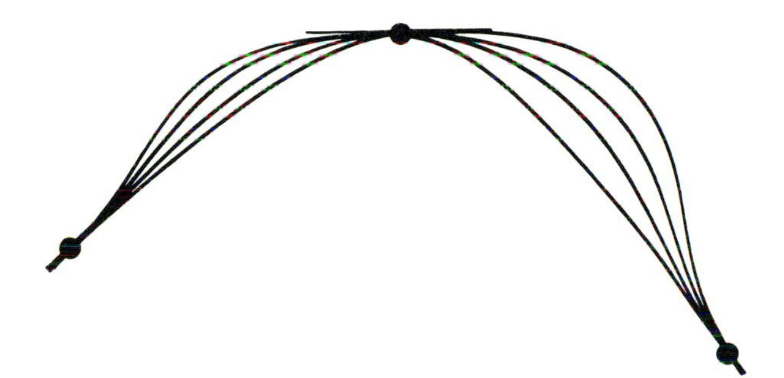

3) As the order of curves becomes higher, undesired oscillations - waviness - will tend to occur. Fifth and sixth order curves may be regarded as a conservative limit.

5.5 Parametric Spline Curves

When too many points or slope constraints must be met for a single polynomial to be used several polynomials may be joined end to end. This can be done by dividing the data points into groups or *spans* and interpolating different polynomials over each span with the constraint that the slopes at the joins should match. The joins are called *knots*. All the simultaneous equations needed to find all the coefficients of all the parametric polynomials are consequently inter-dependent, and the whole linear system needs to be solved in one, rather complicated, operation. The completed structure is called a *spline*. As each span will generally contain few points high order polynomials are not often used in this technique. Despite this, waviness may still occur, and there have been a number of sophisticated spline formulations that attempt to overcome such difficulties.

A much simpler technique, which has a number of advantages, is the parametric *Overhauser* curve. This is generated by dividing the set of points through which the curve is to pass (which we will assume to be fairly regularly spaced) into overlapping groups of three. A parametric quadratic is fitted to each set as outlined in Section 5.4.

A linear blending function is then used to combine the sets of curves into a single curve through all the data points. This blending function has the value one at the middle of each quadratic and zero at the ends. The resulting curve is smooth, and will not exhibit waviness, however many points are to be interpolated.

Looking in more detail at part of the diagram above, suppose that two sets of three points, J K L and K L M are parameterised uniformly as follows:

Span 1, J K L:

At J $s = 0$

At K $s = 0.5$

At L $s = 1$

Span 2, K L M:

At K $t = 0$

At L $t = 0.5$

At M $t = 1$

Suppose also that solving for the constant terms in the parametric quadratics gives the equations:

For J K L

$$x = a_{11} + b_{11}s + c_{11}s^2$$

$$y = a_{12} + b_{12}s + c_{12}s^2$$

and for K L M

$$x = a_{21} + b_{21}t + c_{21}t^2$$

$$y = a_{22} + b_{22}t + c_{22}t^2$$

Then any point on the curve *between K and L* has coordinates

$$x = f(a_{11} + b_{11}s + c_{11}s^2) + g(a_{21} + b_{21}t + c_{21}t^2)$$

$$y = f(a_{12} + b_{12}s + c_{12}s^2) + g(a_{22} + b_{22}t + c_{22}t^2)$$

where the parameters are linked

$$s = t + 0.5$$

and f and g are the linear blending multipliers:

$$f = 2 - 2s = 1 - 2t$$

$g = 2s - 1 = 2t$

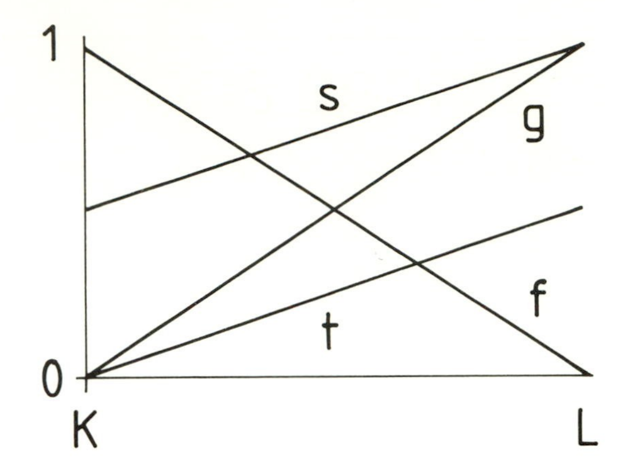

The curve between the first and second points, and between the last and last but one points is not, of course, blended in the parametric Overhauser scheme, but is the unaltered quadratic curve in each case.

5.6 Radius of Curvature

The radius of curvature, R, of an arbitrary curve, $f(x,y) = 0$, at some point on the curve (x,y) is the radius of a circle that would have the same curvature as that of the curve at that point. As this radius goes to infinity at points of inflection it is not as useful as curvature, κ, its inverse:

$$\kappa = \frac{1}{R}$$

κ is given by

$$\kappa = \frac{f_{xx} f_y - 2 f_{xy} f_x f_y + f_{yy} f_x}{[f_x^2 + f_y^2]^{3/2}}$$

where f_{xx} is the second partial derivative of the function with respect to x and so on. For parametric curves, $x = f(t)$, $y = g(t)$, κ is given by:

$$\kappa = \frac{f_t g_{tt} - g_t f_{tt}}{[f_t^2 + g_t^2]^{3/2}}$$

6

Vectors, matrices and transformations

6.1 Vectors

Up to this point we have dealt with geometrical problems in the plane by using explicit pairs of Cartesian coordinates such as x and y to represent points. We have treated x and y as numbers and manipulated them using algebra, or calculated values for them in computer programs. However, it is often simpler (especially in three or more dimensions) to use vector methods to solve geometrical problems. Those who are not familiar with vectors are recommended to consult the book by Macbeath. We shall limit ourselves to a brief list of the properties of vectors and the functions that are used to manipulate them.

A vector is a list of numbers. These might, for example, be coordinates in three dimensions, so a vector, **p**, representing a point in space would have three components:

$$\mathbf{p} = (x_p, y_p, z_p)$$

p is in bold type to show that it is not a number, but a vector. **p** can be considered to be a straight line drawn from the origin of coordinates to the point. The length or *magnitude* of **p** is written as $|\mathbf{p}|$. It is easily seen from Pythagoras' theorem that:

$$|\mathbf{p}| = \sqrt{(x_p^2 + y_p^2 + z_p^2)}$$

Vectors do not need to be considered as always starting from the origin. For example if we know that another point is x_q away from **p** in the x direction, y_q in the y direction and z_q in the z direction then we can construct a vector, **q**:

$$\mathbf{q} = (x_q, y_q, z_q)$$

and we can say that the position of the second point relative to the origin of coordinates, as opposed to relative to **p**, is given by the vector **s**, where:

$$\mathbf{s} = \mathbf{p} + \mathbf{q} = (x_p, y_p, z_p) + (x_q, y_q, z_q)$$

$$= (x_p + x_q, \; y_p + y_q, \; z_p + z_q)$$

So to add (and subtract) vectors we just add (and subtract) their individual components. Multiplying a vector by a simple number (a scalar) does not change its direction, but it does change its magnitude (length). We just multiply all the elements of the vector by the scalar:

$$2\mathbf{p} = 2(x_p, y_p, z_p) = (2x_p, 2y_p, 2z_p)$$

There are two ways of multiplying one vector by another. First we shall consider the scalar, dot or inner product of two vectors. All three terms are synonymous, and are in general use. From here on we shall call this product the scalar product.

If our two vectors, **p** and **q**, have an angle between them of θ, then we define the scalar product, **p.q**, which is a single number, as:

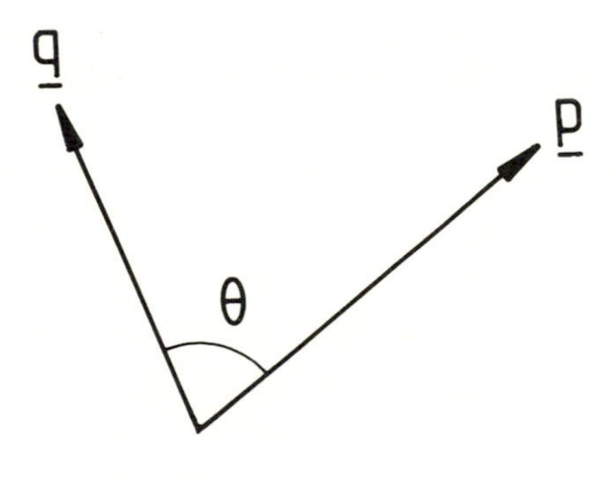

$$\mathbf{p.q} = |\mathbf{p}||\mathbf{q}|\cos\theta$$

As can be seen from elementary trigonometry the scalar product is equal to the magnitude of one of the vectors times the length of the projection of the other onto it. The scalar product is also equal to the sum of the products of the components of the vectors:

$$\mathbf{p.q} = (x_p, y_p, z_p).(x_q, y_q, z_q)$$

$$= x_p x_q + y_p y_q + z_p z_q$$

In most problems we know the components of the vectors, but not the angle between them, so this is the more useful form of the scalar product for doing calculations. The ability of the scalar product to project any vector onto another with only a few additions and multiplications is very useful, and we will return to it later.

The second method of multiplying vectors is known as the vector product, and the result of this operation is another vector. The vector product of **p** and **q**, **pxq**, is defined as

$$\mathbf{p}\times\mathbf{q} = \mathbf{n}\ |\mathbf{p}|\,|\mathbf{q}|\sin\theta$$

where **n** is a vector of unit magnitude (length) at right angles to the plane containing **p** and **q** such that the three form a right-handed coordinate system.

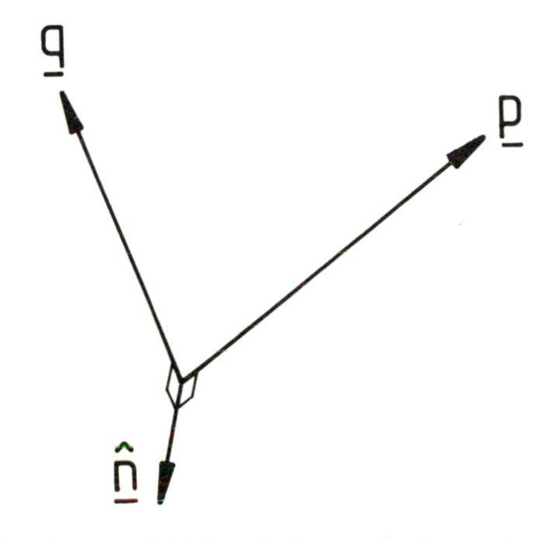

p, **q** and **n** forming a right handed coordinate system

As can be seen the magnitude of **pxq** is twice the area of the triangle lying between the vectors **p** and **q**, and the vector **pxq** must lie in a direction at right angles to the two vectors that we started with. This ability of the vector product to produce vectors at right angles to any others is useful, and we will return to that later as well. As with the scalar product it is possible to write an expression for the vector product that only uses the components of the two vectors to be multiplied and does not involve θ. This expression is the determinant (see Section 6.3)

$$\mathbf{p}\times\mathbf{q} = \begin{vmatrix} \mathbf{i} & \mathbf{j} & \mathbf{k} \\ x_p & y_p & z_p \\ x_q & y_q & z_q \end{vmatrix}$$

which expands to

$$\mathbf{p}\times\mathbf{q} = (y_p z_q - y_q z_p)\mathbf{i} - (x_p z_q - x_q z_p)\mathbf{j} + (x_p y_q - x_q y_p)\mathbf{k}$$

where **i**, **j**, and **k** are three vectors with a magnitude of 1.0 (*unit vectors*) in the direction of the x, y, and z axes respectively. The result of evaluating this determinant is an expression involving the sum of **i**, **j** and **k** all multiplied by coefficients. As these are all unit vectors in the coordinate directions the resulting vector simply has as its x component the coefficient of **i**, as its y component the coefficient of **j**, and, as its z component, the coefficient of **k**.

Most computer languages require that calculations involving vectors are done explicitly on the elements of those vectors, so, once the algebra of a geometrical problem has been solved, the vectors used have to be decomposed into their components when the solution is being coded into a program. However, it is still worth using vector methods to solve problems at the pencil and paper stage, as they tend greatly to simplify the working, and make it less likely that mistakes will be made. Some computer languages (for example APL) allow the programmer to use vectors and matrices (see below) directly, and thus are a powerful tool for the solving of geometrical problems. Their only drawback is that they tend not to be widely available.

6.2 Matrices

A matrix is a rectangle of numbers. Matrices are represented by heavy capital letters, for example **A**, or a rectangle of numbers in round brackets:

$$\mathbf{A} = \begin{pmatrix} a_{11} & a_{12} \\ a_{21} & a_{22} \end{pmatrix}$$

This is a two by two matrix. A one by n matrix is just a vector, of course. Readers unfamiliar with matrices are recommended to read the books by Stevenson. Again, a detailed description of matrices and the functions that operate on them is outside the scope of this book, but it is worth mentioning the multiplication of a vector by a matrix. If **p** is a vector with two components, written as a column vector

$$\mathbf{p} = \begin{pmatrix} x_p \\ y_p \end{pmatrix}$$

we can write:

$$\mathbf{Ap} = \begin{pmatrix} a_{11} & a_{12} \\ a_{21} & a_{22} \end{pmatrix} \begin{pmatrix} x_p \\ y_p \end{pmatrix} = \begin{pmatrix} a_{11}x_p + a_{12}y_p \\ a_{21}x_p + a_{22}y_p \end{pmatrix}$$

The result on the right is another column vector.

6.3 Determinants

A determinant is just a number. Determinants arise from the consideration of linear simultaneous equations (for example the need to find the intersection between two straight lines - see Chapter 1, Section 1.5) when the coefficients of the variables in the equations have to be multiplied and added in a fixed, symmetrical manner in order to obtain a solution. If you are not familiar with determinants you are recommended to consult Stevenson's books. Here we shall restrict ourselves to a brief description of how a determinant is written down, and how it is calculated.

Determinants are written as a square of numbers surrounded by vertical bars, for example:

$$DETERM = \begin{vmatrix} d_{11} & d_{12} & d_{13} \\ d_{21} & d_{22} & d_{23} \\ d_{31} & d_{32} & d_{33} \end{vmatrix}$$

This is a third order determinant, as it has three rows and columns. A determinant is evaluated by scanning along one of its rows or columns and alternately adding and subtracting the value of the determinant formed by omitting the row and column corresponding to the value multiplied by that value. Thus, scanning along the top row, we have:

$$DETERM = d_{11}(d_{22}d_{33} - d_{32}d_{23}) -$$

$$d_{12}(d_{21}d_{33} - d_{31}d_{23}) +$$

$$d_{13}(d_{21}d_{32} - d_{31}d_{22})$$

The terms in brackets are the determinants formed by (in order) the bottom right four elements, the left elements on the bottom two rows and the right elements on the bottom two rows, and the bottom four left elements. The following code is a function for evaluating three by three determinants. For a more general routine see, for example, the book by Wilkinson and Reinsch. The elements of the determinant are supplied to the function in the 3x3 array D.

```
FUNCTION DETERM(D)
DIMENSION D(3,3)
        S11 = D(2,2)*D(3,3) - D(3,2)*D(2,3)
        S12 = D(2,1)*D(3,3) - D(3,1)*D(2,3)
        S13 = D(2,1)*D(3,2) - D(3,1)*D(2,2)
        DETERM = D(1,1)*S11 - D(1,2)*S12 + D(1,3)*S13
RETURN
END
```

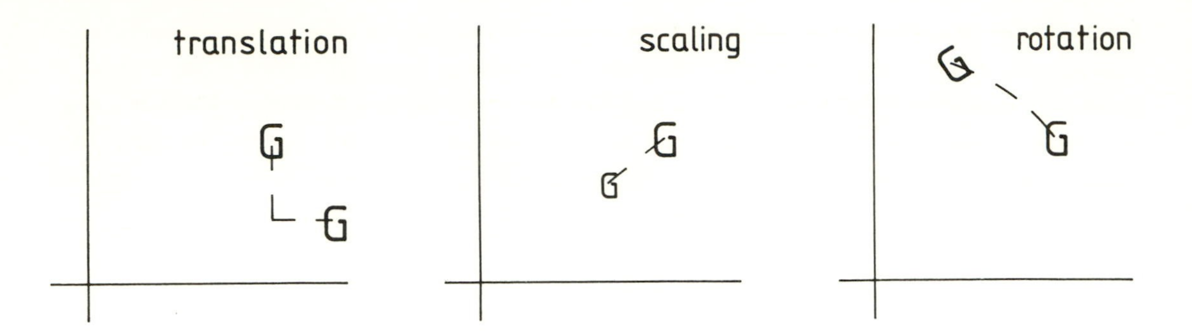

Suppose that **p** is a point in the plane with coordinates x_p and y_p and that we wish to rotate it about the origin of coordinates through an angle θ. The resulting position of **p**, **p'**, will be given by:

$$\mathbf{p'} = \begin{pmatrix} \cos\theta & -\sin\theta \\ \sin\theta & \cos\theta \end{pmatrix} \begin{pmatrix} x_p \\ y_p \end{pmatrix} = \begin{pmatrix} x_p\cos\theta - y_p\sin\theta \\ x_p\sin\theta + y_p\cos\theta \end{pmatrix}$$

This rotation is simply coded in a program:

```
C = COS(THETA)
S = SQRT(1.0 - C*C)
XPP = XP*C - YP*S
YPP = XP*S + YP*C
```

We can translate **p** to a new position by adding another vector, **q** say, to **p**. We can change scales by multiplying **p** by a constant. Using combinations of these three transformations, it is possible to move a point (or a collection of points) about in the plane. The scale change and the rotation can be combined by making the elements of the matrix equal to the scaling factor times the sine and cosine of the required angle.

When changing the scale of a pattern of points by, say, multiplying by 2, the points will also move twice as far away from the origin. Also, when we rotate a pattern of points through an angle θ, the rotation swings about the origin, so the pattern of points is moved somewhere else in space, as well as being rotated. In some circumstances this is undesirable, and it can be avoided by translating the pattern so that its centroid is at the origin, then applying the rotation and scale change, then translating the points back again. If it is also desired to move the pattern this translation can be added on at the final stage.

If we have a pattern of n points, $\mathbf{p}_1, \mathbf{p}_2 .... \mathbf{p}_n$, their centroid will be at a point $\mathbf{p}_c$ given by:

$$\mathbf{p}_c = \frac{1}{n}\sum_{i=1}^{i=n} \mathbf{p}_i$$

All these operations can be accomplished simultaneously using the following code:

```
      DIMENSION P(2,N)              Array for the point components

                .
                .
                .

      XPC = 0.0
      YPC = 0.0

      DO 10 I = 1,N
              XPC = XPC + P(1,I)
              YPC = YPC + P(2,I)    Work out the centroid, pc
10    CONTINUE

      PNINV = 1.0/FLOAT(N)
      XPC = XPC*PNINV
      YPC = YPC*PNINV

      CSCA = COS(THETA)*SCALE
      SSCA = SIN(THETA)*SCALE
      XPCXQ = XPC + XQ              q is the translation vector
      YPCYQ = YPC + YQ


                                    Apply the transformation

      DO 20 I = 1, N
              XPTEM = P(1,I) - XPC
```

```
            YPTEM = P(2,I) - YPC
            P(1,I) = XPTEM*CSCA - YPTEM*SSCA + XPCXQ
            P(2,I) = XPTEM*SSCA + YPTEM*CSCA + YPCYQ
   20    CONTINUE
```

All these operations can be made to work in three dimensions, so it becomes a relatively simple matter to take a three dimensional object defined by some points in space and move it about. Matrix transformation is much more general than even this implies, as, by changing the elements of the matrix, it is possible to shear objects as well as changing their scales, and to project them from three dimensions down into two (in other words to ensure that the third component of the vectors that result from the matrix multiplications is always zero).

Where many transformations are to be performed one after another there is a technique using 4 by 4 matrices called *homogeneous coordinates* which allows all three transformations to be done by matrix multiplications. Homogeneous coordinates are rather inefficient on ordinary computers (those without matrix manipulation hardware). They are, however, elegant, and this has encouraged other authors to use them exclusively. Newman and Sproull, and Faux and Pratt, give good treatments.

6.5 Perspective

One of the most common requirements of computer graphics programs is the production of a two-dimensional picture of a three-dimensional object. The three dimensional object will usually be stored in the form of points in space representing its corners. These may possibly be collected together in records to define faces of the object, and so on. Data structures for storing such objects will not be covered here - if you are interested in this problem (which has received a great deal of attention from computer scientists) you are recommended to consult Newman and Sproull.

The problem that we will consider is how to take a point in three dimensions, **p**, and find the point on a graphics screen in two dimensions, **q**, which is its two dimensional 'image'. Once this problem has been solved, producing a picture from the points becomes a relatively easy matter. Any straight lines in three dimensions that may join the points will transform to straight lines in two dimensions, so only the endpoints need to be transformed.

Another problem that we will not consider is how to remove the hidden detail from such a picture (the parts of the object being depicted that lie behind other parts and are obscured by them). Again, a great deal of work has been done on this problem, which is surprisingly difficult to solve. This is also dealt with in Newman and Sproull.

The first, and simplest, form of perspective that we will consider is general isometric projection. This will be familiar to readers who have studied engineering drawing. In this form of perspective two lines are drawn at angles α and β to the horizontal.

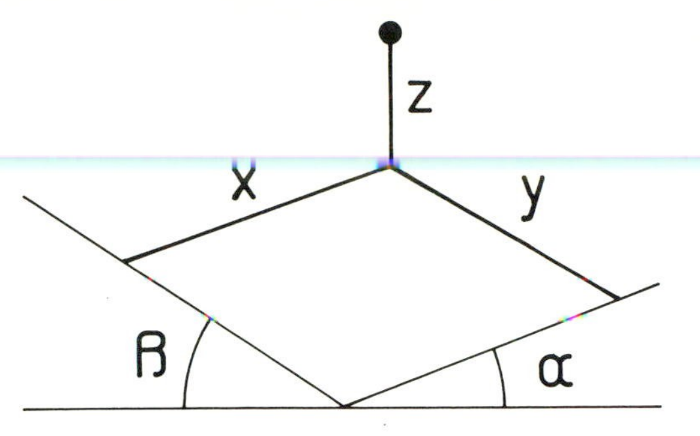

The x components of the points to be transformed are measured along the right hand line, and the y components along the left. The z component is then added vertically to the corner of the resulting parallelogram. As can be seen (and as the name of the perspective transformation implies) this process preserves the values of the coordinates from three dimensions down into the two-dimensional image. In standard isometric perspective α and β are both 30°, but it is possible to simulate different viewing positions by changing the two angles.

To obtain **q** from **p** we multiply **p** by the matrix:

$$\begin{pmatrix} \cos\alpha & -\cos\beta & 0 \\ \sin\alpha & \sin\beta & 1 \\ 0 & 0 & 0 \end{pmatrix}$$

The result will be a vector, **q**, whose third component is always zero (because of the zero bottom line of the matrix). This vector will be the position (relative to the origin of the image) of **p** in two dimensions. As before all that is needed to rescale the data is for the matrix to be multiplied by the desired scaling factor before the transformation is made. By changing the 1 in the right hand column for a number less than 1 the z scale can be compressed by that amount without changing the x and y information. z can be expanded by choosing a number greater than 1. The x information can be compressed or expanded by reducing or increasing the $\cos\alpha$ and $\sin\alpha$ terms by the same multiple while leaving the rest of the matrix unchanged, and the y information can be altered in the same way by changing the $\cos\beta$ and $\sin\beta$ terms.

The following code will perform the transformation. This code is probably best written as a subroutine which can be called to transform a point whenever it is needed, though when several points are being transformed the setting up of the sine and cosine information should only be done once - the last two lines would form the transformation subroutine. The code incorporates three separate scaling factors for the x, y and z directions, SCAX, SCAY and SCAZ. These can be omitted (or set to 1.0) if simple one to one scaling is all that is wanted.

```
CALPHA = COS(ALPHA)*SCAX
SALPHA = SIN(ALPHA)*SCAX
CBETA = COS(BETA)*SCAY              Set up the elements of the matrix
```

```
SBETA = SIN(BETA)*SCAY

UQ = CALPHA*XP - CBETA*YP                  find q
VQ = SALPHA*XP + SBETA*YP + SCAZ*ZP
```

The problem with this sort of perspective is that it does not produce a real life image. In other words the picture is not what would be obtained by pointing a camera at the object being drawn and taking a picture. However, it may be useful to people who wish to write programs to produce engineering drawings. A common misconception is that isometric projection makes the hidden detail removal problem mentioned above easier than it is if true, camera, perspective is used. If the hidden detail part of a program is properly written it should be no less efficient on true perspective than it is on isometric perspective.

The second sort of perspective transformation is true, or camera, perspective. This requires rather more code than isometric perspective, but the actual transformation (the equivalent of the last two lines of the code above) does not require much more code or computer time.

The naming conventions used for vectors in what follows are that the scalar products of pairs of vectors are stored in variables named with the two vector letters (e.g. WE is **w.e**). Vector products are new vectors, and are thus given new variables to store their components in.

What would we need to know in order to produce a perspective projection of a pattern of points in space?

Clearly we would need:

1) The position of the eye in space
2) Which way it was pointing
3) Which way it considered to be 'up'
4) The positions of the points

The two subroutines described below take such information and form an image from it. They transfer information between themselves in a COMMON block called PSPBLK. The first routine, called PSPSET, defines a perspective projection given the information in items 1 to 3 above, and the second, called PSP, transforms a point, **p**, in three dimensions to a point **q** in a two-dimensional image plane.

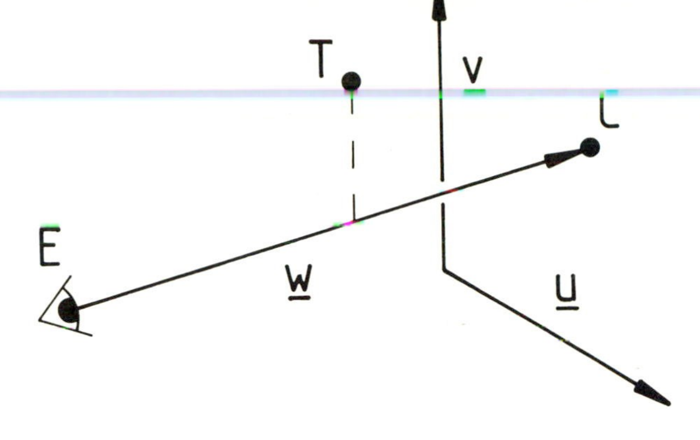


```
SUBROUTINE PSPSET(XE,YE,ZE,XC,YC,ZC,XT,YT,ZT)
COMMON/PSPBLK/ XU,YU,ZU,UE,XV,YV,ZV,VE,XW,YW,ZW,WE
```

(XE, YE, ZE) is the eye position
(XC, YC, ZC) is a point to be in the centre of the required picture.
(XT, YT, ZT) is a point that is to appear above (XC,YC,ZC)

***w** is the line of sight*

```
XW = XC - XE
YW = YC - YE
ZW = ZC - ZE
R = XW*XW + YW*YW + ZW*ZW
IF (R.LT.ACCY) THEN
```

..... *Eye position and centre coincide*

```
ELSE
        RINV = 1.0/SQRT(R)         Normalise w to
        XW = XW*RINV               unit length
        YW = YW*RINV
        ZW = ZW*RINV
ENDIF

WE = XW*XE + YW*YE + ZW*ZE          Project e onto w

XU = YW*(ZT - ZE) - ZW*(YT - YE)     u is at right
YU = ZW*(XT - XE) - XW*(ZT - ZE)     angles to t-e
```

```
      ZU = XW*(YT - YE) - YW*(XT - XE)      and w - it is the
                                            picture's 'x' axis
      R = XU*XU + YU*YU + ZU*ZU
      IF (R.LT.ACCY) THEN

            ..... t coincides with e

      ELSE
            RINV = 1.0/SQRT(R)              Normalise u
            XU = XU*RINV
            YU = YU*RINV
            ZU = ZU*RINV
      ENDIF

      UE = XU*XE + YU*YE + ZU*ZE            Project e onto u

      XV = YU*ZW - ZU*YW                    v is at right
      YV = ZU*XW - XU*ZW                    angles to u and
      ZV = XU*YW - YU*XW                    w - it is the
                                            picture's 'y' axis

      VE = XV*XE + YV*YE + ZV*ZE            Project e onto v
RETURN
END
```

The result of calling this routine is three vectors, **u**, **v**, and **w**, and three scalar products, UE, VE, and WE. **u** is at right angles to the line of sight, **w**, and to **t** - **e** (as it is their vector product), so it must correspond to the 'x' axis of the picture. UE is the projection of the eye position onto that axis. **v** is at right angles to **u** and to the line of sight, so it must be the picture's 'y' axis. This vector does not need to be normalised, as it is the vector product of two unit vectors. VE is the projection of the eye position onto that 'y' axis. **w** is the line of sight, and WE is the projection of the eye position onto that. To perform a perspective transformation of a point, $\mathbf{p} = (x_p, y_p, z_p)$, to a point in two dimensions, **q**, we use the routine:

```
SUBROUTINE PSP(XP,YP,ZP,UQ,VQ)
COMMON/PSPBLK/ XU,YU,ZU,UE,XV,YV,ZV,VE,XW,YW,ZW,WE

      The output position is q = (u_q, v_q)

      R = XW*XP + YW*YP + ZW*ZP - WE
      IF (R.LT.ACCY) THEN

            ..... Point is behind the eye
```

```
            ELSE
                      RINV = 1.0/R
                      UQ = RINV*(XU*XP + YU*YP + ZU*ZP - UE)
                      VQ = RINV*(XV*XP + YV*YP + ZV*ZP - VE)
            ENDIF
      RETURN
      END
```

This projects the point to be transformed onto the unit vectors representing the picture's 'x' and 'y' axes and subtracts the projection of the eye position to get the absolute position. The multiplication by RINV scales the picture so that far away objects are reduced in size appropriately. If it is omitted the projection will be parallel (in other words as seen from infinity). The image produced is in a plane through the point **c** perpendicular to the line of sight (**w** = **c** - **e**). Points in space are joined to the eye, and the coordinates returned (UQ, VQ) are the point where these sight lines cut the image plane, with the centre of view, **c**, as the origin.

Generally the simplest way to use these routines is to compute the eight corners of the smallest cuboid that completely encloses the object to be plotted (in other words the range of x, y, and z values for the object) and set up a transformation with PSPSET, looking from an eye position outside that cuboid. Next find the values of **q** corresponding to thse eight corners, and work out the range of u and v coordinates required to accommodate the whole image on the plotting device. The values of **q** returned for the actual object being plotted can then be scaled so as to fit in.

It is easy to produce stereoscopic image pairs using these routines - just generate an image, move the eye position slightly, and generate another.

7

Points, lines and planes

7.1 Distance Between Two Points in Space

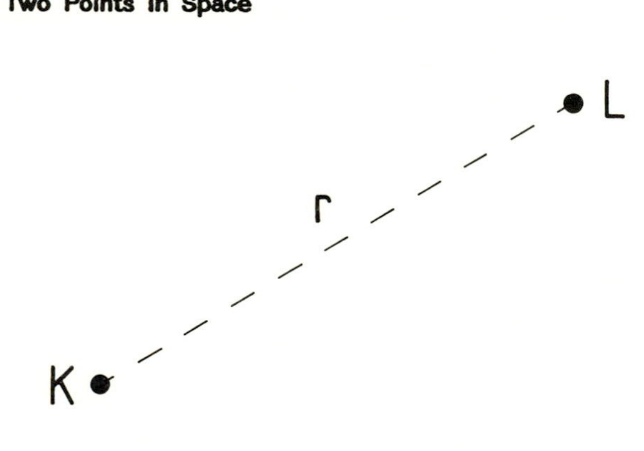

Given two points in space (x_K, y_K, z_K) and (x_L, y_L, z_L) the distance between them, r, is given by Pythagoras theorem as in the two dimensional case described in Section 1.1:

$$r = \sqrt{[(x_L - x_K)^2 + (y_L - y_K)^2 + (z_L - z_K)^2]}$$

All the computational considerations mentioned in Section 1.1 apply in the three-dimensional case as well. In particular it may often be more convenient and quicker to encode an algorithm using squared distances throughout a data structure than to waste time using the expensive square root function.

In three dimensions, the implicit equation of a line is the intersection of two planes. Since more than one pair of planes can describe a given line, the parametric form is to be preferred. This is very similar to the two-dimensional parametric equation of a line:

$$x = x_0 + ft$$

$$y = y_0 + gt$$

$$z = z_0 + ht$$

This may be specified so that the parameter, t, has values 0 and 1 at the ends of a segment (see Section 7.6). Alternatively, the normalised form may be prefered, and, as in two dimensions, this means that a change in t corresponds to real distance moved along the line. When the equations are normalised the condition

$$f^2 + g^2 + h^2 = 1$$

must be satisfied. This can be achieved by dividing f, g, and h by $\sqrt{(f^2 + g^2 + h^2)}$ to obtain new values:

```
DENSQ = F*F + G*G + H*H
IF (DENSQ.LT.ACCY) THEN

          ..... The parametric coefficients are corrupted

ELSE
          DINV = 1.0/SQRT(DENSQ)
          F = F*DINV
          G = G*DINV
          H = H*DINV
ENDIF
```

In the normalised form the coefficients f, g, and h are the cosines of the angles the line makes with the coordinate axes.

The remaining arbitrariness in these equations can be removed by making the point (x_0, y_0, z_0) the point on the line nearest to the origin of coordinates. This is the point where the normal from the origin meets the line. This may avoid numerical problems with points (x_0, y_0, z_0) that are very distant from the origin, and it may facilitate comparisons of several lines, especially to detect coincident lines. Note that an accuracy constant must be used when comparing floating point numbers.

The normal from from the origin to a parametric line meets the line where

$$t = \frac{-(fx_0 + gy_0 + hz_0)}{(f^2 + g^2 + h^2)}$$

but when the line is in its normalised form the bottom line may be omitted, as it is 1. The t = 0 point of a normalised line may be moved to the point nearest the origin as follows:

```
D = F*XO + G*YO + H*ZO
XO = XO - F*D
YO = YO - G*D
ZO = ZO - H*D
```

7.3 Distance from a Point to a Line in Space

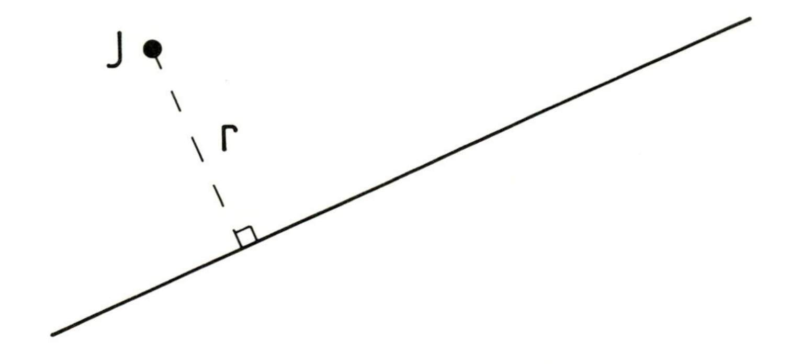

If the line equation is

$$x = x_0 + ft$$

$$y = y_0 + gt$$

$$z = z_0 + ht$$

then the value of the parameter, t, at the point on the line nearest to the given point (x_J, y_J, z_J) is:

$$t = \frac{f(x_J - x_O) + g(y_J - y_O) + h(z_J - z_O)}{(f^2 + g^2 + h^2)}$$

If the line equation is normalised the denominator is 1 of course. If we let $x_{JO} = (x_J - x_O)$, and so on, then the squared distance from the point to the line, r^2, is given by:

$$r^2 = \frac{[g(fy_{JO} - gx_{JO}) + h(fz_{JO} - hx_{JO})]^2 + [f(gx_{JO} - fy_{JO}) + h(gz_{JO} - hy_{JO})]^2 + [f(hx_{JO} - fz_{JO}) + g(hy_{JO} - gz_{JO})]^2}{(f^2 + g^2 + h^2)^2}$$

This can be coded:

```
DENOM = F*F + G*G + H*H
IF (DENOM.LT.ACCY) THEN

        ..... The line parameter coefficients are corrupt

ELSE
        XJO = XJ - XO
        YJO = YJ - YO
        ZJO = ZJ - ZO

        FYGX = F*YJO - G*XJO
        FZHX = F*ZJO - H*XJO
        GZHY = G*ZJO - H*YJO

        V1 = G*FYGX + H*FZHX
        V2 = H*GZHY - F*FYGX
        V3 = - F*FZHX - G*GZHY

        R = SQRT(V1*V1 + V2*V2 + V3*V3)/DENOM
ENDIF
```

7.4 Distance Between Two Lines in Space

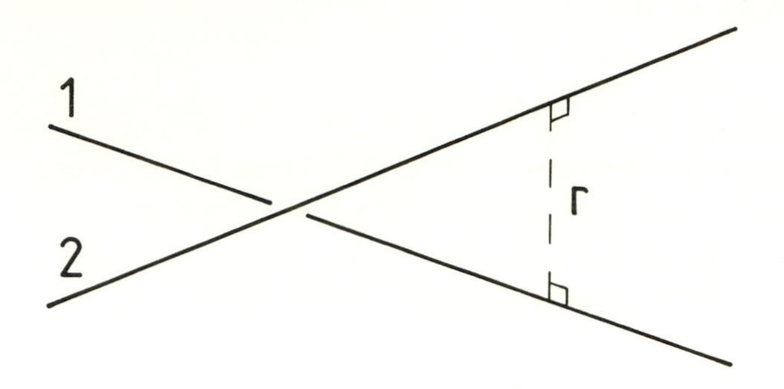

If the two lines have parametric equations

$$x = x_1 + f_1 s$$

$$y = y_1 + g_1 s$$

$$z = z_1 + h_1 s$$

and

$$x = x_2 + f_2 t$$

$$y = y_2 + g_2 t$$

$$z = z_2 + h_2 t$$

then the minimum distance between them is given by the expression:

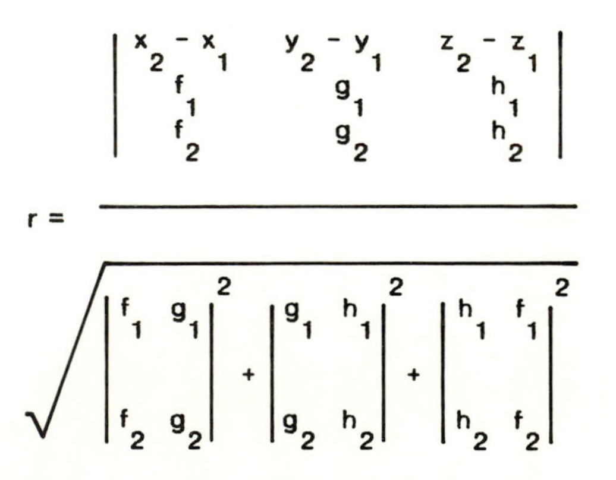

$$r = \frac{\begin{vmatrix} x_2 - x_1 & y_2 - y_1 & z_2 - z_1 \\ f_1 & g_1 & h_1 \\ f_2 & g_2 & h_2 \end{vmatrix}}{\sqrt{\begin{vmatrix} f_1 & g_1 \\ f_2 & g_2 \end{vmatrix}^2 + \begin{vmatrix} g_1 & h_1 \\ g_2 & h_2 \end{vmatrix}^2 + \begin{vmatrix} h_1 & f_1 \\ h_2 & f_2 \end{vmatrix}^2}}$$

Considerable exploitation of subexpressions may be made in coding this, which then becomes:

```
X21 = X2 - X1
```

```
Y21 = Y2 - Y1
Z21 = Z2 - Z1

FG = F1*G2 - F2*G1
GH = G1*H2 - G2*H1
HF = H1*F2 - H2*F1

DENOM = FG*FG + GH*GH + HF*HF
IF (DENOM.LT.ACCY) THEN

        ..... The lines are parallel

ELSE
        R = ABS(X21*GH + Y21*HF + Z21*FG)/SQRT(DENOM)
ENDIF
```

The sign of the expression is an indication of the relative parametric direction of the two lines. This information is much more easily obtained in other ways (see Section 7.5) and so the sign is discarded.

Note that this code does not work for parallel and near-parallel lines. The geometry in such cases is inherently unstable, and it is best to reformulate the problem itself as one of finding the distance from a *point* to a line (see Section 7.3).

7.5 Angle Between Two Lines in Space

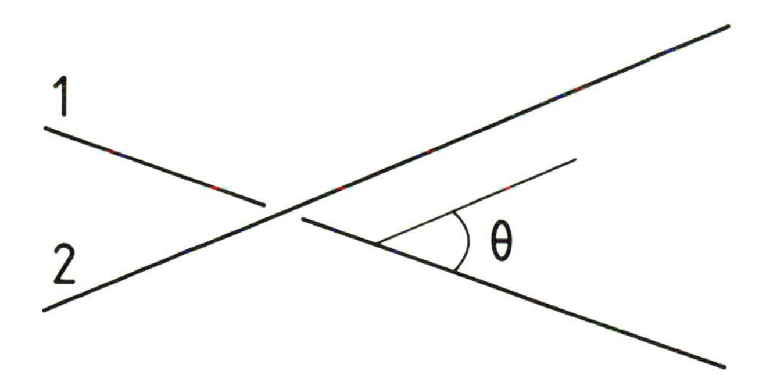

The angle between two lines is found from the scalar product of their direction vectors; it is not necessary for the lines to intersect. For the two lines

$$x = x_1 + f_1 s$$

$$y = y_1 + g_1 s$$

$$z = z_1 + h_1 s$$

and

$$x = x_2 + f_2 t$$

$$y = y_2 + g_2 t$$

$$z = z_2 + h_2 t$$

with normalised coefficients, the angle between them is given by:

$$\theta = \cos^{-1}(f_1 f_2 + g_1 g_2 + h_1 h_2)$$

If the lines are not normalised, then using the expression

$$\theta = \cos^{-1} \frac{f_1 f_2 + g_1 g_2 + h_1 h_2}{\sqrt{[(f_1^2 + g_1^2 + h_1^2)(f_2^2 + g_2^2 + h_2^2)]}}$$

avoids one square root operation and several divisions when compared with normalising the lines separately before finding the angle between them. This is coded:

```
DENOM = (F1*F1 + G1*G1 + H1*H1)*(F2*F2 + G2*G2 + H2*H2)
IF (DENOM.LT.ACCY) THEN
          ..... One or both lines have corrupted coefficients
ELSE
          THETA = ACOS((F1*F2 + G1*G2 + H1*H2)/SQRT(DENOM))
ENDIF
```

The ACOS function returns values in the range 0 to π.

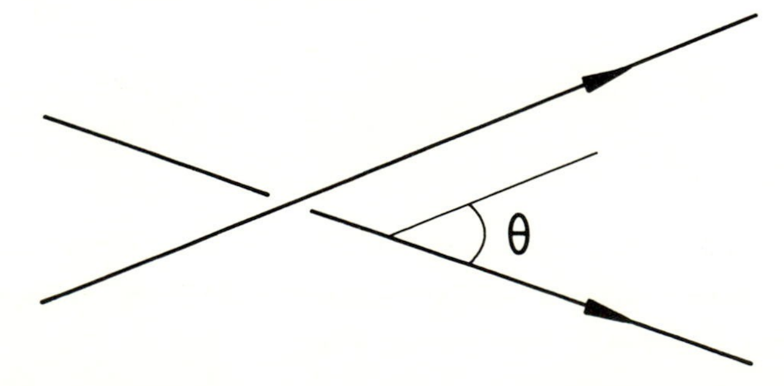

The angle measured is that between the lines with parameters moving in the same direction. $\theta = 0$ indicates that the lines are parallel in the same direction, and $\theta = \pi$ that they are parallel in the opposite direction. Note that rounding error makes it unlikely that results for parallel lines will give exactly 0 or exactly π.

If the acute angle is required values over $\pi/2$ must be subtracted from π.

For old FORTRAN compilers that do not have an ACOS function, an alternative using ATAN2 is described in Section 1.4.

7.6 Line through Two Points in Space

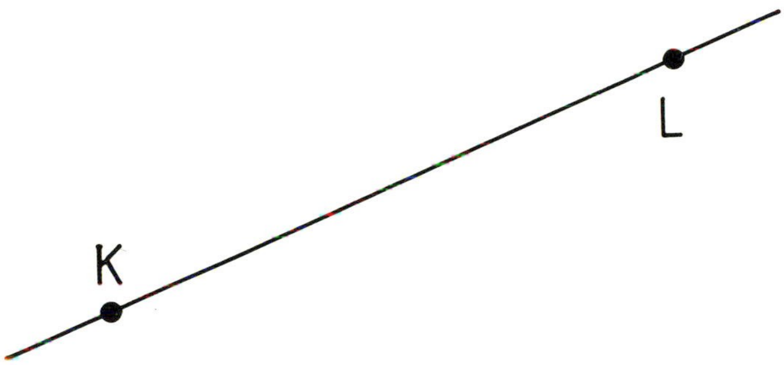

K

L

The specification of a parametric line in space given two points on it is a direct extension of the two-dimensional formula given in Section 1.6. Again the parameter value, t, goes from 0 at the first point to 1 at the second. The line equations are:

$$x = x_K + (x_L - x_K)t$$

$$y = y_K + (y_L - y_K)t$$

$$z = z_K + (z_L - z_K)t$$

This is not normalised unless the distance between the points is 1. It can be normalised by dividing the coefficients of t (but *not* the constant terms) by the square root of sum of squares of those coefficients.

7.7 Equation of a Plane

Both the implicit and the parametric equations of planes in space are useful. The implicit form is simpler in general, and more compact. The parametric form is to be preferred when essentially two-dimensional operations are to be performed *in* the plane, because it readily permits axes to be defined in it.

The Implicit Form

This is the direct equivalent of the implicit line equation in two dimensions, and is:

$$ax + by + cz + d = 0$$

In the normalised form $a^2 + b^2 + c^2 = 1$. To convert an unnormalised implicit plane equation to its normalised form multiply the entire equation by:

$$\frac{1}{\sqrt{(a^2 + b^2 + c^2)}}$$

In the normalised form a, b, and c are the cosines of the angles which the normal to the plane makes with the coordinate axes. The absolute value of d is the perpendicular distance of the origin from the plane. The distance between two parallel normalised planes, 1 and 2, may therefore be determined as $|d_2 - d_1|$.

As with the implicit line equation, a normalised plane equation may be multiplied through by −1. The plane may be considered to represent the boundary of a semi-infinite region of space by applying a convention that the vector formed by the direction cosines (a, b, and c) always points towards the outside (or the inside) of the region. With such a convention, the equation is that of a planar half-space, a boundary between a volume of solid and an empty volume. Such half-spaces can be used to define convex polyhedra and more complicated three-dimensional shapes and regions in space by using the set theoretic methods described in Section 4.5 for two dimensions.

The Parametric Form

A single parameter may be used to specify distance along a line. Two are required to specify a position in a plane. Given a point in space (x_0, y_0, z_0), and two *different* vectors (i.e. directions) both parallel to the plane (f_1, g_1, h_1) and (f_2, g_2, h_2), a point in the plane is found by adding a proportion of one vector and a different proportion of the second vector to the point coordinates:

$$x = x_0 + f_1 s + f_2 t$$

$$y = y_0 + g_1 s + g_2 t$$

$$z = z_0 + h_1 s + h_2 t$$

If the two vectors have unit length, and are perpendicular, that is the conditions

$$f_1^2 + g_1^2 + h_1^2 = 1$$

$$f_2^2 + g_2^2 + h_2^2 = 1$$

$$f_1 f_2 + g_1 g_2 + h_1 h_2 = 0 \qquad \text{(scalar product = zero)}$$

are fulfilled, then the parameters s and t constitute measurements along orthogonal axes in the plane from an origin in it (x_0, y_0, z_0), and two-dimensional geometry of the type described in the first five chapters of this book may be performed in the plane by using s and t in place of x and y. We have, in fact, already met this in the section on true perspective projection. The final stages of the projection were two-dimensional geometry in an image plane perpendicular to the direction in which an eye was pointing (see Section 6.4).

Conversion Between Implicit and Parametric Planes

As with the parametric line, the point on the plane nearest to the origin of coordinates is often a convenient one to use as the origin of the parametric system. It corresponds to the parameter values

$$s = \frac{(f_2 x_0 + g_2 y_0 + h_2 z_0)(f_1 f_2 + g_1 g_2 + h_1 h_2) - (f_1 x_0 + g_1 y_0 + h_1 z_0)(f_2^2 + g_2^2 + h_2^2)}{(f_1^2 + g_1^2 + h_1^2)(f_2^2 + g_2^2 + h_2^2) - (f_1 f_2 + g_1 g_2 + h_1 h_2)^2}$$

$$t = \frac{(f_1 x_0 + g_1 y_0 + h_1 z_0)(f_1 f_2 + g_1 g_2 + h_1 h_2) - (f_2 x_0 + g_2 y_0 + h_2 z_0)(f_1^2 + g_1^2 + h_1^2)}{(f_1^2 + g_1^2 + h_1^2)(f_2^2 + g_2^2 + h_2^2) - (f_1 f_2 + g_1 g_2 + h_1 h_2)^2}$$

which can, of course, be much simplified if the vectors defining the parametric system were normalised. If the point corresponding to these parameter values is (x_0', y_0', z_0'), then the implicit plane equation $ax + by + cz + d = 0$ is readily derived in its normalised form:

$$d = \sqrt{(x_0'^2 + y_0'^2 + z_0'^2)}$$

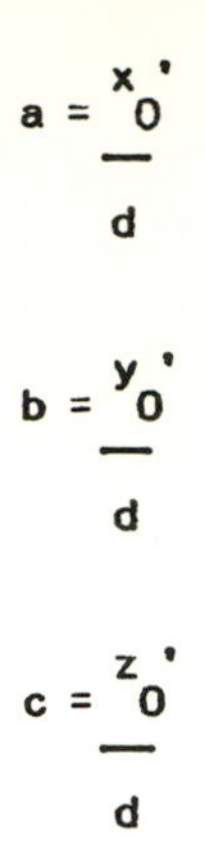

$$a = \frac{x_0'}{d}$$

$$b = \frac{y_0'}{d}$$

$$c = \frac{z_0'}{d}$$

Conversely, to go from implicit to parametric equations, we can write down (x_0, y_0, z_0) from the plane equation (assumed not to be normalised):

$$x_0 = \frac{da}{\sqrt{(a^2 + b^2 + c^2)}}$$

$$y_0 = \frac{db}{\sqrt{(a^2 + b^2 + c^2)}}$$

$$z_0 = \frac{dc}{\sqrt{(a^2 + b^2 + c^2)}}$$

Selecting the vector system for the parametric equation is somewhat arbitrary, however. We may choose a first vector for external reasons, and the second is then formed from the vector product of that vector and the normal vector (a, b, c). If we have no particular interest in the choice of the vector system, then the first vector can be created as the vector product of the normal and any arbitrary vector, which must, of course, make a reasonably large angle with it. The coordinate axis corresponding to the smallest of a, b, or c is a convenient choice.

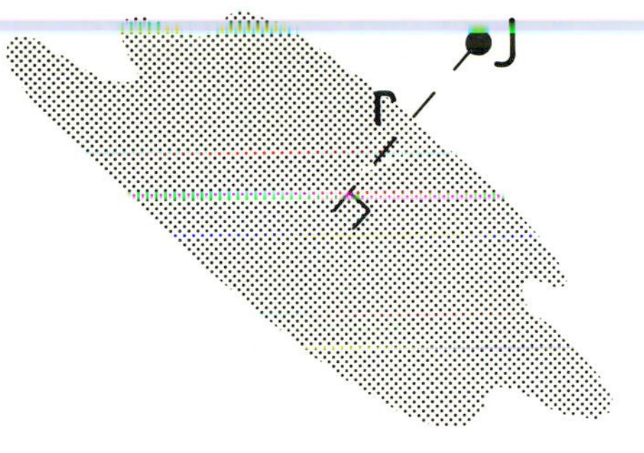

If the plane is $ax + by + cz + d = 0$ and the point is (x_J, y_J, z_J) then the square of the perpendicular distance from the point to the plane, r^2, is:

$$r^2 = \frac{(ax_J + by_J + cz_J + d)^2}{a^2 + b^2 + c^2}$$

This is very similar to the two-dimensional formula given in Section 1.3. The code is:

```
DENOM = A*A + B*B + C*C
IF (DENOM.LT.ACCY) THEN

            ..... The plane equation is corrupt

ELSE
            SR = A*XJ + B*YJ + C*ZJ + D
            RSQ = SR*SR/DENOM
ENDIF
```

If the plane is normalised, the single statement

```
SR = A*XJ + B*YJ + C*ZJ + D
```

suffices. As in Section 1.3, a positive value of SR indicates that the point is on the side of the plane towards which the vector (a, b, c) is pointing. Negative values indicate that it is on the other side. If this information is irrelevant the absolute value of SR should be taken.

7.9 Angle between a Line and a Plane

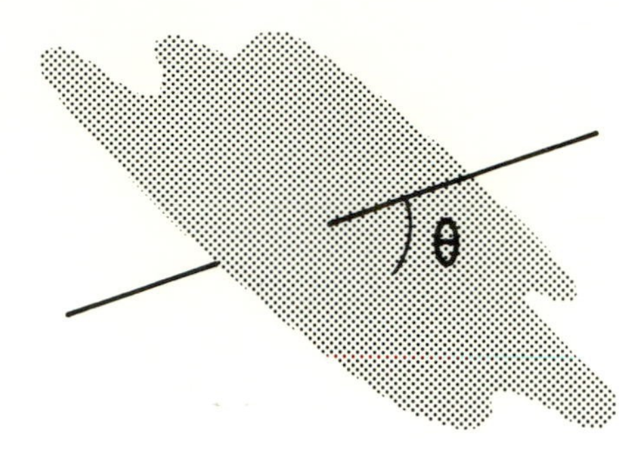

The angle between a line and a plane is the angle between the normal to the plane and the line, subtracted from a right angle. The first angle, between the line and the normal to the plane, may be found, as usual, from a scalar product.

If the plane is

$$ax + by + cz + d = 0$$

and the line is

$$x = x_0 + ft$$

$$y = y_0 + gt$$

$$z = z_0 + ht$$

the angle between the line and normal is

$$\gamma = \cos^{-1}(af + bg + ch)$$

if both line and plane are normalised.

The following expressions

$$\gamma = \cos^{-1} \frac{af + bg + ch}{\sqrt{(a^2 + b^2 + c^2)}}$$

$$\gamma = \cos^{-1} \frac{af + bg + ch}{\sqrt{(f^2 + g^2 + h^2)}}$$

and

$$\gamma = \cos^{-1} \frac{af + bg + ch}{\sqrt{[(a^2 + b^2 + c^2)(f^2 + g^2 + h^2)]}}$$

may be used when just the line, just the plane, and neither are normalised; they are efficient alternatives to local normalisation of the line and plane. The angle between the line and the plane is then:

$$\theta = \frac{\pi}{2} - \gamma$$

This should be in the range $-\pi/2$ to $\pi/2$. Positive values indicate that the line's parameter moves away from the plane on the same side as its normal points. Negative values indicate the reverse, and if $\theta = 0$, of course, the line is parallel to the plane.

The most general solution is coded:

```
DENOM = (A*A + B*B + C*C)*(F*F + G*G + H*H)
IF (DENOM.LT.ACCY) THEN

          ..... Either or both equations are corrupt

ELSE
          SPROD = A*F + B*G + C*H
          GAMMA = ACOS(SPROD/SQRT(DENOM))
          THETA = 1.570796 - GAMMA
ENDIF
```

7.10 Angle between Two Planes

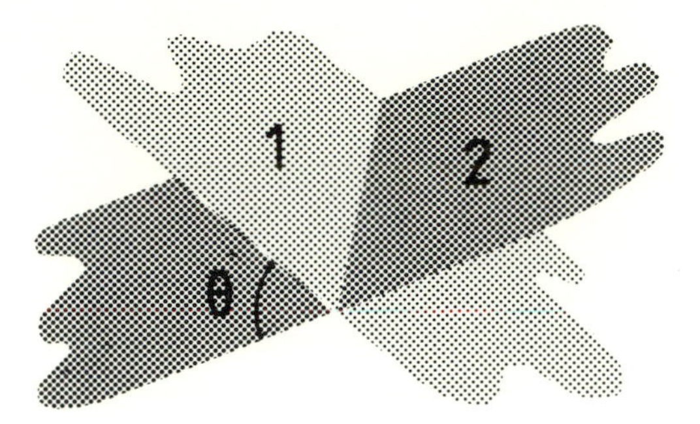

This problem is very similar to that solved in the previous section. To find the angle between two planes we find the angle between their normals. If the two planes are

$$a_1x + b_1y + c_1z + d_1 = 0$$

and $$a_2x + b_2y + c_2z + d_2 = 0$$

then, if the coefficients are normalised, the angle between the planes is

$$\theta = \cos^{-1}(a_1a_2 + b_1b_2 + c_1c_2)$$

and with unnormalised coefficients it is:

$$\theta = \cos^{-1}\frac{a_1a_2 + b_1b_2 + c_1c_2}{\sqrt{[(a_1^2 + b_1^2 + c_1^2)(a_2^2 + b_2^2 + c_2^2)]}}$$

This latter equation will code to run faster than separately normalising the planes and then finding the angle between them. The code needed is identical to that used in the last section, and results are again in the range $0 - \pi$. When the plane equations represent half-spaces, the angle is determined considering them in the same sense, ie the solid semi-infinite spaces would coincide if one of the half-spaces were to be rotated through θ.

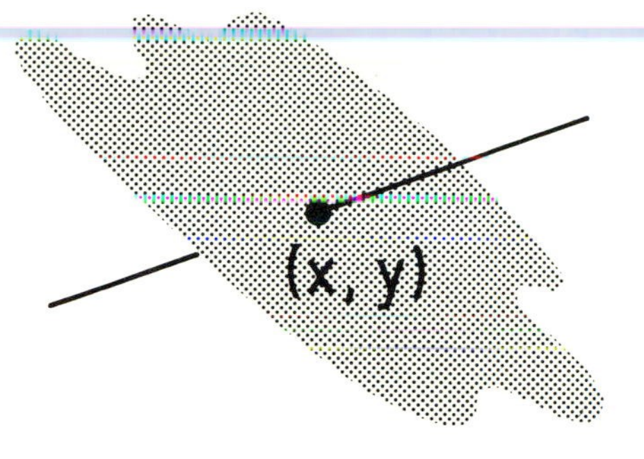

If the plane is

$$ax + by + cz + d = 0$$

and the line is

$$x = x_0 + ft$$

$$y = y_0 + gt$$

$$z = z_0 + ht$$

then the parameter on the line where it intersects the plane is given by

$$t = \frac{-(ax_0 + by_0 + cz_0 + d)}{af + bg + ch}$$

and the intersection point can be found using the code:

```
DENOM = A*F + B*G + C*H
IF (ABS(DENOM).LT.ACCY) THEN

        ..... The line and plane are parallel

ELSE
        T = -(A*XO + B*YO + C*ZO + D)/DENOM
        X = XO + F*T
        Y = YO + G*T
        Z = ZO + H*T
```

```
ENDIF
```

This code is completely general - none of the equations need to be normalised.

7.12 Intersection of Three Planes

This problem is the three-dimensional equivalent of the two-dimensional problem of finding the point where two straight lines intersect. As before the solution is to treat the plane equations as three simultaneous linear equations and solve them. If the planes are

$$a_1x + b_1y + c_1z + d_1 = 0$$

$$a_2x + b_2y + c_2z + d_2 = 0$$

$$a_3x + b_3y + c_3z + d_3 = 0$$

then the following code will find the intersection point (x, y, z):

```
BC = B2*C3 - B3*C2
AC = A2*C3 - A3*C2
AB = A2*B3 - A3*B2
DET = A1*BC - B1*AC + C1*AB
IF (ABS(DET).LT.ACCY) THEN

          ..... At least two planes are parallel

ELSE
          DC = D2*C3 - D3*C2
          DB = D2*B3 - D3*B2
```

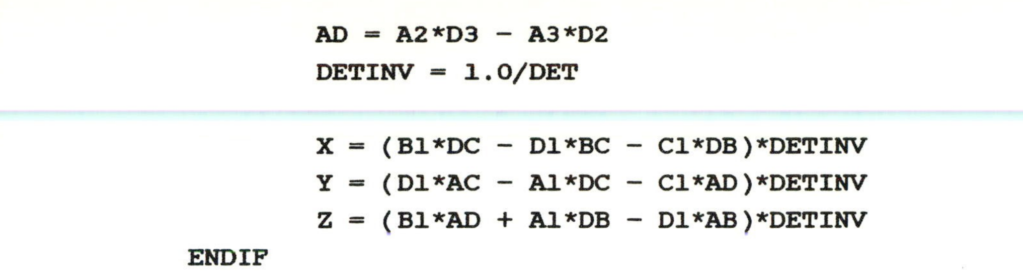

```
            AD = A2*D3 - A3*D2
            DETINV = 1.0/DET

            X = (B1*DC - D1*BC - C1*DB)*DETINV
            Y = (D1*AC - A1*DC - C1*AD)*DETINV
            Z = (B1*AD + A1*DB - D1*AB)*DETINV
      ENDIF
```

7.13 Intersection of Two Planes

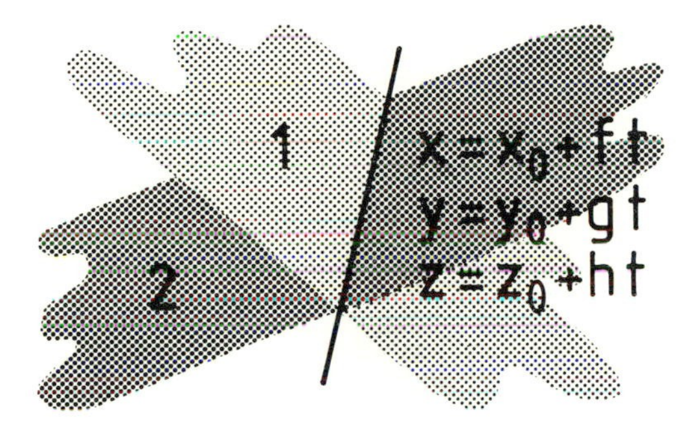

Two planes intersect to form a straight line. If the planes are

$$a_1x + b_1y + c_1z + d_1 = 0$$

and $$a_2x + b_2y + c_2z + d_2 = 0$$

and the line of intersection that we want to find is

$$x = x_0 + ft$$

$$y = y_0 + gt$$

$$z = z_0 + ht$$

then the parameter coefficients, f, g, and h, are given by:

$$f = \begin{vmatrix} b_1 & c_1 \\ b_2 & c_2 \end{vmatrix} \qquad g = \begin{vmatrix} c_1 & a_1 \\ c_2 & a_2 \end{vmatrix} \qquad h = \begin{vmatrix} a_1 & b_1 \\ a_2 & b_2 \end{vmatrix}$$

As this stands f, g, and h are not normalised. If we choose to make the point (x_0, y_0, z_0) the point on the line of intersection nearest to the origin then the following code will find the line:

```
F = B1*C2 - B2*C1
G = C1*A2 - C2*A1
H = A1*B2 - A2*B1
DET = F*F + G*G + H*H
IF (DET.LT.ACCY) THEN

        ..... The planes are parallel

ELSE
        DETINV = 1.0/DET
        DC = D1*C2 - C1*D2
        DB = D1*B2 - B1*D2
        AD = A1*D2 - A2*D1

        XO = (G*DC - H*DB)*DETINV
        YO = -(F*DC + H*AD)*DETINV
        ZO = (F*DB + G*AD)*DETINV
ENDIF
```

If the reader wishes to normalise the line equation he should include code in the ELSE clause to multiply F, G, and H by SQRT(DETINV), but this should not work out the square root three times.

7.14 Plane through Three Points

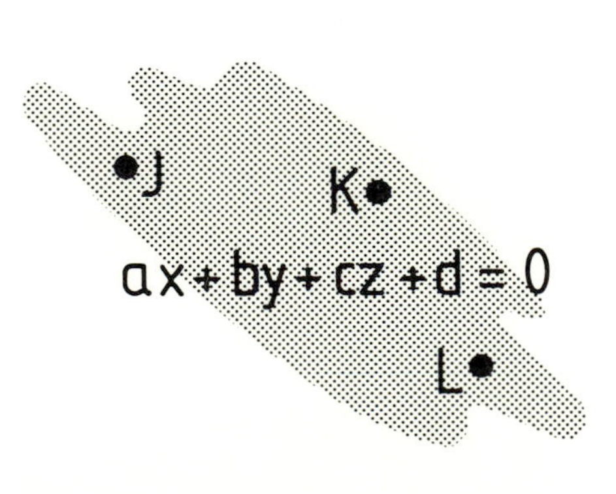

The implicit equation of a plane through three points may be stated as a determinant containing the three independent variables x, y and z

$$\begin{vmatrix} x - x_J & y - y_J & z - z_J \\ x_K - x_J & y_K - y_J & z_K - z_J \\ x_L - x_J & y_L - y_J & z_L - z_J \end{vmatrix} = 0$$

Which effectively states that a vector formed by J and any point in the plane must be perpendicular to the vector product of the vectors from J to K and from J to L. This vector product is, of course, normal to the plane, so the variable vector from J must lie in the plane.

A normalised form of the plane equation is not usually produced, and the numerical accuracy of the code below relies on the points being spaced well apart and not all lying close to a straight line (in other words the nearer they are to forming an equilateral triangle the better).

If the determinant above is multiplied out it gives the usual form of the plane equation:

$$ax + by + cz + d = 0$$

When coding this considerable exploitation of repeated expressions can be made:

```
XKJ = XK - XJ
YKJ = YK - YJ
ZKJ = ZK - ZJ
XLJ = XL - XJ
YLJ = YL - YJ
ZLJ = ZL - ZJ

A = YKJ*ZLJ - ZKJ*YLJ
B = ZKJ*XLJ - XKJ*ZLJ
C = XKJ*YLJ - YKJ*XLJ
D = -(XK*A + YK*B + ZK*C)
```

7.15 Plane through a Point and Normal to a Line

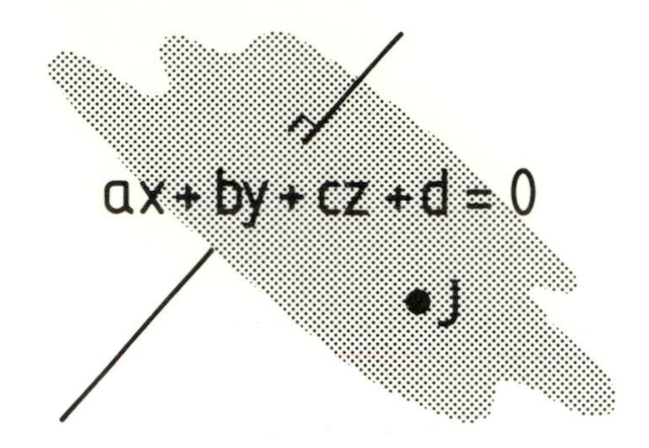

This is not a difficult calculation, because a plane is described in terms of its normal. If the line is

$$x = x_0 + ft$$

$$y = y_0 + gt$$

$$z = z_0 + ht$$

then the plane normal to that line passing through a point J is

$$fx + gy + hz - (fx_J + gy_J + hz_J) = 0$$

The plane will only be normalised if the line was.

7.16 Plane through Two Points and Parallel to a Line

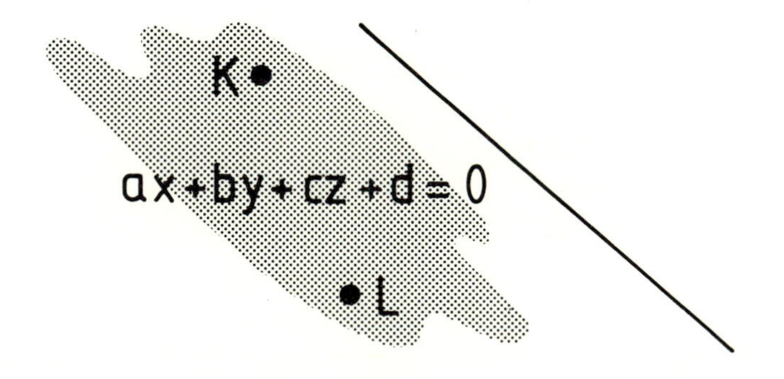

If the given line is

$$x = x_0 + ft$$

$$y = y_0 + gt$$

$$z = z_0 + ht$$

then the plane parallel to it which passes through K and L is:

$$(hy_{LK} - gz_{LK})x + (fz_{LK} - hx_{LK})y + (gx_{LK} - fy_{LK})z -$$
$$x_K(hy_{LK} - gz_{LK}) - y_K(fz_{LK} - hx_{LK}) - z_K(gx_{LK} - fy_{LK}) = 0$$

where $x_{LK} = x_L - x_K$, $y_{LK} = y_L - y_K$, and $z_{LK} = z_L - z_K$

This, despite its complexity, is not even normalised. However, the number of repeated terms means that the code is relatively simple, and that it is also easy to produce a normalised answer. If we consider the above equation to reduce to

$$ax + by + cz + d = 0$$

then to get the normalised coefficient in this equation we may use the code:

```
XLK = XL - XK
YLK = YL - YK
ZLK = ZL - ZK
A = H*YLK - G*ZLK
B = F*ZLK - H*XLK
C = G*XLK - F*YLK
DENOM = A*A + B*B + C*C
IF (DENOM.LT.ACCY) THEN

          ..... Points coincident or line coefficients corrupt

ELSE
          DENINV = 1.0/SQRT(DENOM)
          A = A*DENINV
          B = B*DENINV
          C = C*DENINV
          D = -(A*XK + B*YK + C*ZK)
ENDIF
```

Note that the point (x_0, y_0, z_0) is not needed to solve this problem.

8
Volumes

8.1 Volume of a Tetrahedron

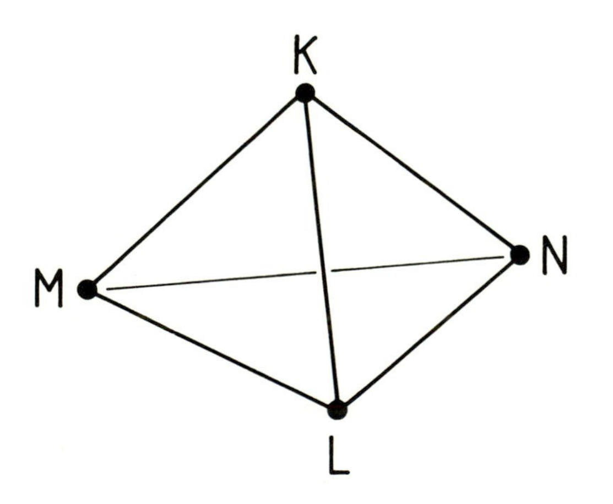

This volume is particularly useful because it is often possible to split more complicated solids into tetrahedra and to add up the volumes of those tetrahedra to find the total volume of the solid.

The volume of a tetrahedron is found by forming a determinant from the three vectors obtained by subtracting one vertex (we choose K) from the other three and then dividing by 6:

$$\frac{1}{6}\begin{vmatrix} (x_L - x_K) & (x_M - x_K) & (x_N - x_K) \\ (y_L - y_K) & (y_M - y_K) & (y_N - y_K) \\ (z_L - z_K) & (z_M - z_K) & (z_N - z_K) \end{vmatrix}$$

This is simple to code:

```
XLK = XL - XK
YLK = YL - YK
ZLK = ZL - ZK

XMK = XM - XK
YMK = YM - YK
ZMK = ZM - ZK

XNK = XN - XK
YNK = YN - YK
ZNK = ZN - ZK

DMN = YMK*ZNK - ZMK*YNK
DLN = YLK*ZNK - ZLK*YNK
DLM = YLK*ZMK - ZLK*YMK

VOL = 0.166667*(XLK*DMN - XMK*DLN + XNK*DLM)
VOL = ABS(VOL)
```

If the absolute value of the volume is not taken, its sign indicates the relationship between vertex point K (in general the subtracted point) and the other three points. If the determinant is positive L, M, and N appear in clockwise order when seen from K; if negative anti-clockwise. This may be used to decide which side of a triangle (which may be representing part of a surface) a point lies. Alternatively a collection of points can be categorised as lying on one side or the other of the plane through three of the points by calculating values of VOL (there is no need to divide by 6 for this, of course) and categorising using the sign of the result.

This method of calculating the volume of a tetrahedron is directly analogous to the similar method used to calculate the area of a triangle in Section 4.1.

8.2 Centre of Gravity and Surface Area of a Tetrahedron

The centroid of a tetrahedron is also useful; it is given by

$$x_{CG} = 0.25(x_K + x_L + x_M + x_N)$$

$$y_{CG} = 0.25(y_K + y_L + y_M + y_N)$$

$$z_{CG} = 0.25(z_K + z_L + z_M + z_N)$$

which, like the volume, is a direct extension of the corresponding formula for the triangle.

The surface area of a tetrahedron is, of course, the sum of the areas of its four triangular faces. The area of a triangle in space can be calculated directly from the vector product of the relative positions of its vertices. For example for the triangle K L M:

$$a_{KLM} = \frac{1}{2}\sqrt{\{ [(y_L - y_K)(z_M - z_K) - (z_L - z_K)(y_M - y_K)]^2 + [(z_L - z_K)(x_M - x_K) - (x_L - x_K)(z_M - z_K)]^2 + [(x_L - x_K)(y_M - y_K) - (y_L - y_K)(x_M - x_K)]^2 \}}$$

This can be compactly coded, as each position relative to vertex K appears twice, and need only be calculated once.

8.3 Circumcentre of a Tetrahedron

The algebra that gives the centre and radius of the sphere that passes through the four vertices of a tetrahedron is rather complicated, and is not reproduced here. Essentially what is done is to find the determinant formed by subtracting one vertex from the other three in just the same way as the volume of a tetrahedron was found (Section 8.1) and then to substitute a vector of squared distances from the subtracted vertex to the other three into each column of the determinant in turn for each coordinate direction. This second determinant is divided by twice the original one to obtain the position of the centre of the sphere relative to the subtracted vertex. The squared radius of the sphere is found by summing the squares of this relative position, which is then made absolute by adding it to the subtracted vertex. This is very similar to the method employed in Section 4.4

The code to find the circumcentre and squared radius is:

```
XLK = XL - XK          Positions of vertices
YLK = YL - YK          relative to vertex K
ZLK = ZL - ZK
```

```
      XMK = XM - XK
      YMK = YM - YK
      ZMK = ZM - ZK

      XNK = XN - XK
      YNK = YN - YK
      ZNK = ZN - ZK

      DYZ = YMK*ZNK - YNK*ZMK               Sub determinants
      DXZ = XMK*ZNK - XNK*ZMK
      DXY = XMK*YNK - XNK*YMK

      DET = XLK*DYZ - YLK*DXZ + ZLK*DXY     Volume determinant
      IF (ABS(DET).LT.ACCY) THEN

            ..... All four points lie in a plane

      ELSE
            DETINV = 0.5/DET

            RLKSQ = XLK*XLK + YLK*YLK + ZLK*ZLK   Squared distances
            RMKSQ = XMK*XMK + YMK*YMK + ZMK*ZMK   from vertex K
            RNKSQ = XNK*XNK + YNK*YNK + ZNK*ZNK

            DRX = RMKSQ*XNK - RNKSQ*XMK       More sub determinants
            DRY = RMKSQ*YNK - RNKSQ*YMK
            DRZ = RMKSQ*ZNK - RNKSQ*ZMK

                                              Centre relative to K

            X = DETINV*(RLKSQ*DYZ - YLK*DRZ + ZLK*DRY)
            Y = DETINV*(XLK*DRZ - RLKSQ*DXZ - ZLK*DRX)
            Z = DETINV*(-XLK*DRY + YLK*DRX - RLKSQ*DXY)

            RSQ = X*X + Y*Y + Z*Z             Squared radius

            X = X + XK                        Absolute centre position
            Y = Y + YK
            Z = Z + ZK
      ENDIF
```

If radius, rather than its square, is needed, set

```
            R = SQRT(RSQ)
```

as the line after the definition of RSQ.

8.4 Volume and Surface Area of a Cylinder and of a Sphere

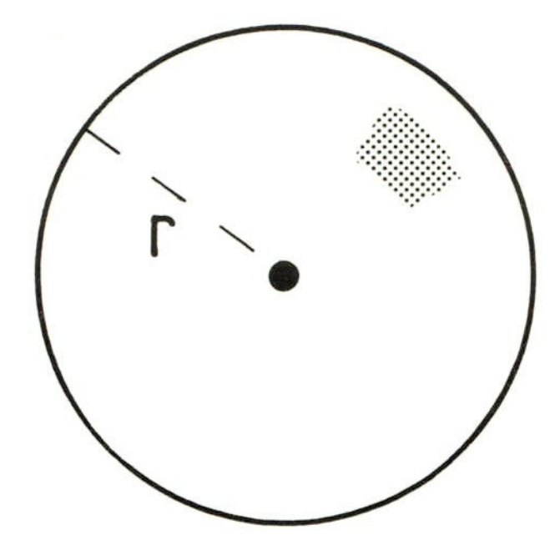

The volume of a sphere is

$$\frac{4}{3}\pi r^3$$

and its surface area is:

$$4\pi r^2$$

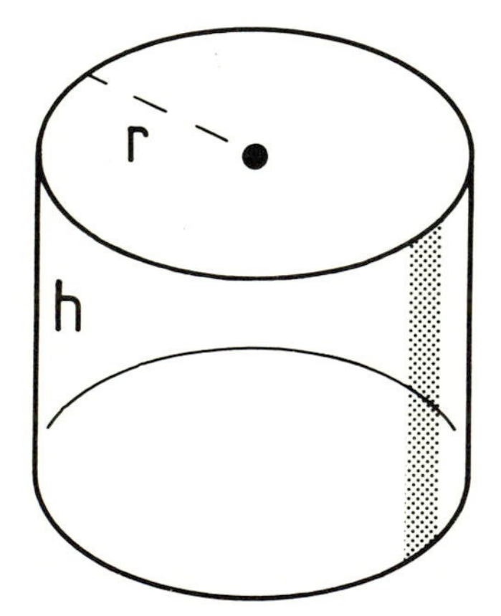

The volume of a cylinder is

$$\pi r^2 h$$

and its surface area (including its ends) is:

$$2\pi r(h + r)$$

It is interesting to note that a cylinder of the same diameter as a sphere, and with a height equal to that diameter, has sides with the same surface area as the sphere (without the circular ends). More than this, if two parallel planes perpendicular to the axis of the cylinder slice the cylinder to make a shorter cylinder, then the area of the sphere sliced out by those planes will equal the area of that short cylinder (again without its ends) as long as both planes intersect the sphere.

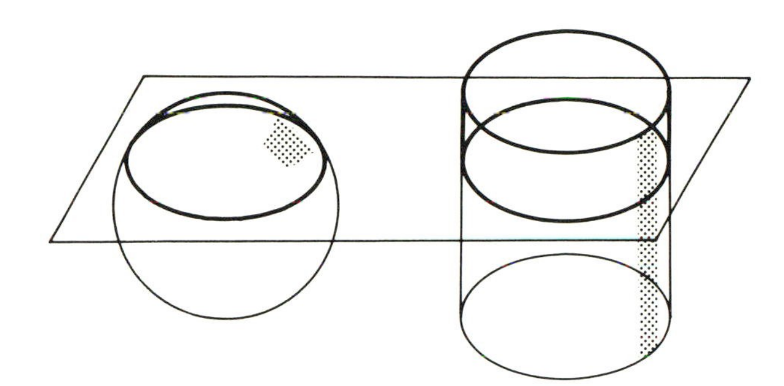

This fact can be used to calculate the area of a sector of a sphere; it is equal to the area of the same length of a cylinder of the same radius.

8.5 Volume and Centre of Gravity of a Sector of a Sphere

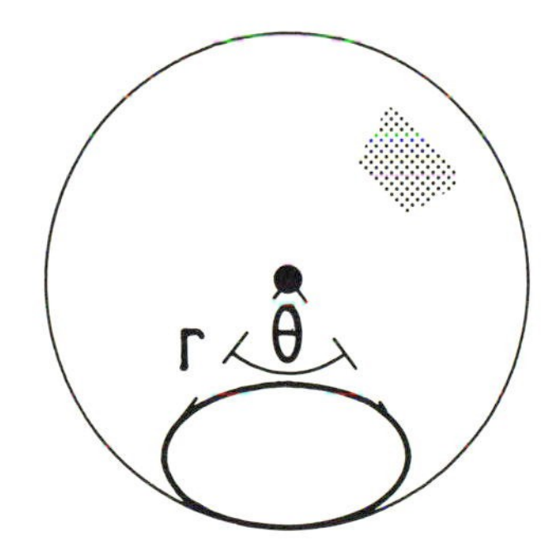

The volume of a sector is

$$\frac{\pi r^3}{3}\left(2 - 3\cos\frac{\theta}{2} + \cos^3\frac{\theta}{2}\right)$$

and the distance from its centroid to the sphere centre is:

$$\frac{3r}{4}\left(\frac{\sin^4\frac{\theta}{2}}{2-3\cos\frac{\theta}{2}+\cos^3\frac{\theta}{2}}\right)$$

The $\sin^4$ term can be replaced

$$\sin^4\frac{\theta}{2} = (1-\cos^2\frac{\theta}{2})^2$$

so that only one call need be made to the trigonometrical functions.

8.6 Volume, Surface Area and Centre of Gravity of a Cone

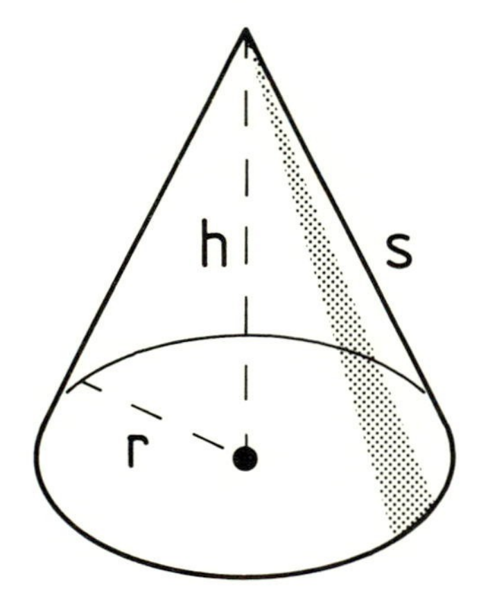

The volume of a cone is:

$$\frac{1}{3}\pi r^2 h$$

Its surface area (including its base) is

$$\pi r(l + r)$$

where l is the slant height of the cone, or

$$\pi r[\sqrt{(h^2 + r^2)} + r]$$

in terms of the height itself. The centroid of a cone is h/4 from its base.

9
Drawing pictures

9.1 Line and Pixel Devices

The purpose of this chapter is to introduce briefly the techniques that are used to produce pictures using computers. The chapter is primarily aimed at those who are building graphics devices for producing such pictures, or who have no graphics software packages available to them, and who consequently have to write their own package starting at the very simplest level.

There are two common mechanisms by which drawings can be produced directly from a computer.

Vector or *calligraphic* devices draw lines directly. Equipment of this sort includes storage tubes, vector refresh displays, and flat-bed and drum pen plotters. These devices are usually limited to a few colours. They can also only draw lines (in some cases of differing thickness), so hatching is needed in order to shade in an area. The highest quality computer line graphics is obtained from large pen plotters, though there are a number of devices available that produce microfilm of almost as good a quality.

The second type of graphics device is the *pixel* display. Here a rectangular matrix of dots forms the computer generated picture. Each dot is called a pixel, which is short for picture element. The pixels may only be able to assume two colours, such as black and white. Examples of such displays are electrostatic plotters and plasma panels. Alternatively the pixels may be able to exhibit a whole range of colours or grey shades to form an image on a raster scan television display. (The word raster is the name given to the zig-zag way in which the electron beam scans across a television screen.)

Deciding how best to use the device available, or what device to choose, depends not only on the actual drawing mechanism, but also on the way in which the picture is described to the device by the computer in use. For instance, most raster scan displays have associated electronics which allow them to decide which pixels to illuminate to construct the closest approximation to a given straight line (see Section 9.4), so the computer need only transmit the end points of the line to the device. The reader will usually find it best to use the facilities provided, especially with devices on slow data transmission lines, when electronics in the display can usually be relied on to do calculations more quickly than they could be transmitted by the driving computer. It is still worth considering carefully how to minimise such things as invisible movement over the plot (i.e. movement with the pen up) on mechanical devices.

In this chapter we describe some of the fundamental techniques that are needed in even the simplest graphics systems. These techniques are described in more detail in a number of other texts, such as Newman and Sproull.

9.2 Circles on Line Devices

We start considering vector devices with a section on circles, as there is clearly no difficulty in drawing straight lines on them.

Some vector devices, such as expensive paper plotters and some vector refresh displays, have circle generation built into them at the hardware level. This is very fast and efficient, and should be used whenever possible.

Otherwise it is necessary to draw a series of line segments, effectively approximating the circle as a polygon. There are three methods of doing this.

The first is to store the vertices of a regular polygon corresponding to a circle of unit radius, and supplying these to the device, appropriately scaled and translated, for each circle required. One immediate problem is that large circles need more segments to appear smooth than small circles. If all circles are generated from one set of stored data then either small ones will take an inordinate time to draw, or large ones will appear coarse. This problem may be solved by storing more than one level of approximation and, possibly, using one of the other two techniques for very large circles. The levels may be interleaved in the same list, so that, for example, to plot a small circle every fourth vertex in the list is used, for larger ones every second vertex is used, and every vertex is used for the largest circles. The memory requirements thus imposed may be minimised by storing only a 45° octant and extrapolating the rest of the circle from that.

Suppose that (x_u, y_u) is a point on the 45° octant of a unit circle. Points on the octants of a circle centre J, radius r_J are then:

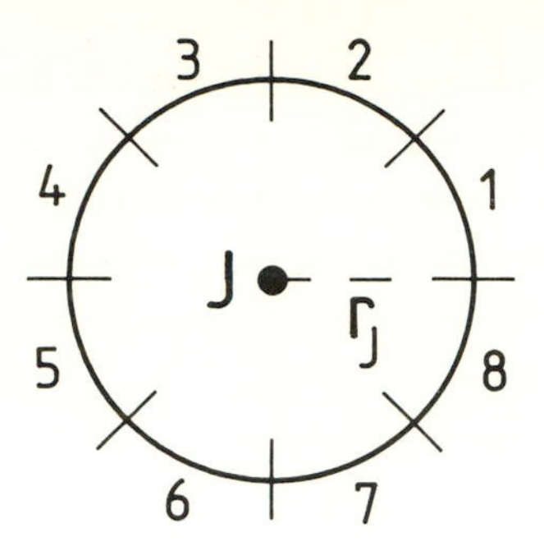

Octant 1

$$x = x_J + r_J x_u$$
$$y = y_J + r_J y_u$$

Octant 2

$$x = x_J + r_J y_u$$
$$y = y_J + r_J x_u$$

Octant 3

$$x = x_J - r_J y_u$$
$$y = y_J + r_J x_u$$

Octant 4

$$x = x_J - r_J x_u$$
$$y = y_J + r_J y_u$$

Octant 5

$$x = x_J - r_J x_u$$
$$y = y_J - r_J y_u$$

Octant 6

$$x = x_J - r_J y_u$$
$$y = y_J - r_J x_u$$

Octant 7

$$x = x_J + r_J y_u$$
$$y = y_J - r_J x_u$$

Octant 8

$$x = x_J + r_J x_u$$
$$y = y_J - r_J y_u$$

If it is wished to move round the circle without reversing direction, the table of unit circle values must be used backwards for even quadrants. This is particularly important on mechanical devices such as pen plotters.

The second technique is simply to interpolate values of θ into the parametric equation of the circle:

$$x = x_J + r_J \cos\theta$$

$$y = y_J + r_J \sin\theta$$

The need to use trigonometric functions for this technique makes it slow, but the line segments are even, and their length is controllable. The number of line segments used may easily be made to vary proportionally with circle radius, remembering that, no matter how small the radius, it is unwise to reduce the number of segments into single figures. When the circle is being drawn on a remote device computation times will probably be negligible compared with the time taken to transmit the coordinates to the device in any case.

Finally the other parametric equations of a circle

$$x = x_J + r_J \frac{(1 - t^2)}{(1 + t^2)}$$

$$y = y_J + r_J \frac{2t}{(1 + t^2)}$$

which have already been discussed in Chapter Two may be used. In this case segments are uneven for equal increments of t. The longest increment will be the first from t = 0 to its first value, and this can be held to a length s by setting the interval of t as:

$$\frac{s}{\sqrt{(4r^2 - s^2)}}$$

9.3 Arcs on Line Devices

Except for special cases of known angle, which may be prestored as in the first circle plotting method, either of the latter two methods of circle parameterisation and plotting may be used for arcs. In both cases the parameter increment is best arranged to give equal steps right up to the arc end, rather than introduce a short last step to accommodate different arc angles, as this can look rather ugly. The last method given for circle plotting above corresponds with the treatment of arcs given in Chapter Three.

Let us assume that we have a subroutine called POINT

```
SUBROUTINE POINT(IX,IY)
```

that sets the pixel at (IX,IY) on the screen of the graphics device. Clearly the screen pixels need to be addressed using integers because of their discrete nature. This routine will not be described, as it will depend entirely on the graphics hardware that the reader is employing.

Let us also assume that we wish to draw a straight line on the device between points (IXJ, IYJ) and (IXK, IYK) on the screen. The following two subroutines will do this pixel by pixel, calling POINT as they go. The algorithm is due to Bressenham.

```
SUBROUTINE LINE(IXJ, IYJ, IXK, IYK)
COMMON /LINBLK/ IXOP, IYOP, NEGX, NEGY, SWPXY
LOGICAL NEGX, NEGY, SWPXY
```

The routine works in one octant and uses
NEGX, NEGY, and SWPXY to move to the other 7

```
        IXOP = IXJ
        IYOP = IYJ
        IXKJ = IXK - IXJ
        IYKJ = IYK - IYJ
        NEGX = (IXKJ.LT.0)
        IF (NEGX) IXKJ = -IXKJ
        NEGY = (IYKJ.LT.0)
        IF (NEGY) IYKJ = -IYKJ
        SWPXY = (IXKJ.LT.IYKJ)
        IF (SWPXY) THEN
                ITEMP = IXKJ
                IXKJ = IYKJ
                IYKJ = ITEMP
        ENDIF
        IX = 0
        IY = 0
        IS = -IXKJ/2

10      IF (IX.LE.IXKJ) THEN
                CALL PLOT(IX,IY)
                IS = IS + IYKJ
                IX = IX + 1
```

```
                    IF (IS.GT.0) THEN
                              IS = IS - IXKJ
                              IY = IY + 1
                    ENDIF
                    GOTO 10
          ENDIF
RETURN
END


SUBROUTINE PLOT (IX,IY)
COMMON /LINBLK/ IXOP, IYOP, NEGX, NEGY, SWPXY
LOGICAL NEGX, NEGY, SWPXY
```

This transforms (IX,IY) to the appropriate quadrant
and sets the pixel there.

```
          IXP = IX
          IYP = IY
          IF(SWPXY) THEN
                    ITEMP = IXP
                    IXP = IYP
                    IYP = ITEMP
          ENDIF
          IF (NEGX) IXP = -IXP
          IF (NEGY) IYP = -IYP
          CALL POINT (IXOP + IXP, IYOP + IYP)
RETURN
END
```

This code always generates lines in the first quadrant, and then rotates them to make them appear in the actual quadrant that they need to be in. This algorithm would almost certainly be implemented in low level code or in hardware. We have simply presented it in FORTRAN for clarity and consistency's sake. The algorithm uses very little arithmetic, so it is well suited to low level devices and code.

9.5 Circles on Pixel Devices

To draw circles on raster scan pixel devices a similar technique is used to the one described in Section 9.4 for drawing straight lines. Suppose that we wish to draw a circle with radius IRJ centred at (IXJ, IYJ). The routine generates a single quadrant of the circle and draws the whole circle by swapping the logical variables used in the quadrant plotting routine, PLOT, of Section 9.4. The algorithm for doing this was invented by Horn.

```
      SUBROUTINE CIRCLE (IXJ, IYJ, IRJ)
      COMMON /LINBLK/ IXOP,IYOP,NEGX,NEGY,SWPXY
      LOGICAL NEGX,NEGY,SWPXY

            NEGX = .TRUE.
            NEGY = .TRUE.
            SWPXY = .TRUE.
            IXOP = IXJ
            IYOP = IYJ
            IX = IRJ
            IY = 0
            IS = -IRJ
10          IF (IX.GE.IY) THEN
                  DO 40 I3 = 1, 2
                        NEGX = (.NOT.NEGX)
                        DO 30 I2 = 1, 2
                              NEGY = (.NOT.NEGY)
                              DO 20 I1 = 1, 2
                                    SWPXY = (.NOT.SWPXY)
                                    CALL PLOT(IX,IY)
20                            CONTINUE
30                      CONTINUE
40                CONTINUE
                  IS = IS + IY + IY + 1
                  IY = IY + 1
                  IF (IS.GT.0) THEN
                        IS = IS - IX - IX + 2
                        IX = IX - 1
                  ENDIF
                  GOTO 10
            ENDIF
      RETURN
      END
```

When plotting pictures on a graphics device it is often useful to be able to magnify part of the plot to fill the whole screen or piece of paper. The rest of the plot then falls off the edges. This process is known as *windowing*, as the plot becomes a window showing part of a larger picture. It is inconvenient to have to write programs so that they work out which parts of the large image will be visible in such cases and only attempt to plot those parts. It is simpler if the program constructs the whole picture, having previously defined the window to the graphics system being used, and the graphics system then cuts out all the invisible detail automatically. This is known as *clipping*. The graphics system may do this clipping by using software on the host computer, or the graphics terminal or plotter may be sufficiently powerful to do the clipping itself. Similar algorithms are used in clipping software in a graphics system and in the electronics of display devices.

We shall describe two clipping algorithms. The first (which is due to Cohen and Sutherland) is for clipping a line to a rectangle which might be the rectangular screen (or part of it) of a graphics device. The second clips a polygon described by an ordered list of its vertices to any convex polygon. Of course this includes rectangles.

Clipping Lines

Consider the problem of clipping a line whose endpoints are (IXK, IYK) and (IXL, IYL) to a rectangle whose bottom left (southwest) corner is the point (IXSW, IYSW) and whose top right (northeast) corner is the point (IXNE, IYNE). We assume that all these values are integer screen coordinates, and that it is necessary to do the clipping using fast integer arithmetic in, say, a microprocessor in a graphics device. To clip a line using floating point arithmetic at a higher level the methods described in Section 3.3 should be used.

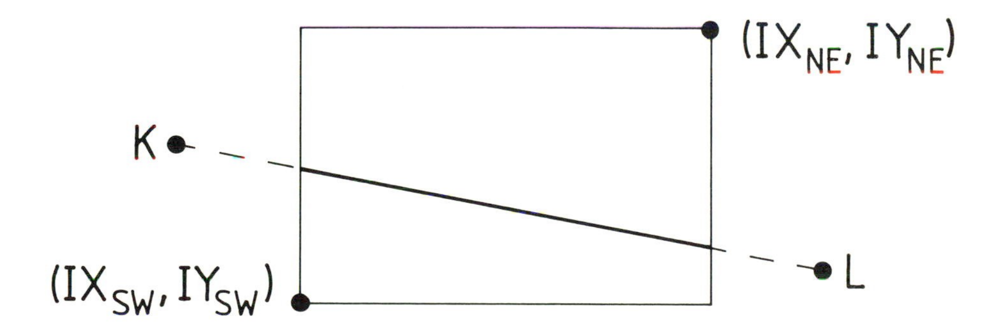

The first step in the algorithm is to categorise both endpoints of the line by a four bit integer (in other words a number between 0 and 15). This categorisation describes the relationship of the points to the clipping rectangle and eight regions round it.

1001 (9)	1000 (8)	1010 (10)
0001 (1)	0000 (0)	0010 (2)
0101 (5)	0100 (4)	0110 (6)

The least significant bit (bit 0 at the right hand end of the number) is set to 1 if the point is to the left of the clipping rectangle. Bit 1 is set if the point is to the right of the rectangle. Bit 2 is set if the point is below the rectangle. Bit 3 is set if the point is above the rectangle. If a point is within the clipping rectangle its category will be zero. If both points have category zero then obviously the whole line segment is visible, and may be plotted immediately. Conversely the line is completely invisible if the set intersection of the bit patterns (logical AND) of the two categories for the line segment's endpoints is not zero, so none of it need be plotted. In either of these cases no clipping calculations need to be performed.

If at least one endpoint category is *not* zero and the result of the AND operation *is* zero, then part of the line segment may intersect the rectangle, and the line must be clipped to find that visible part, if it exists. The clipping is done by repeatedly dividing the line in half to find the points on it that are at the edge of the rectangle. The procedure is very similar to a binary search to find a value in an ordered list - in this case the list is the points along the line. In order to divide the line segment in half the endpoints are added and the sum is divided by two. Integer addition is a fast process, and so is division by two, as that is just a right shift of the bit pattern of a number. The clipping algorithm is applied twice. First to find the furthest visible point from K, second to find the furthest visible point from L. These points are the ends of the visible part of the clipped line segment.

If no visible point is found on the first pass of the algorithm the line is invisible, of course, and no further work needs to be done.

Here is the first pass of the algorithm in FORTRAN terms, though in practice it would usually be implemented in a low level language or in hardware. The function IVIS(IX,IY) returns the category value (0, 1, 2, 4, 5, 6, 8, 9, or 10) of the point (IX,IY). The function IPAND(ICATA,ICATB) returns the result of a logical AND of the two endpoint categories, ICATA and ICATB.

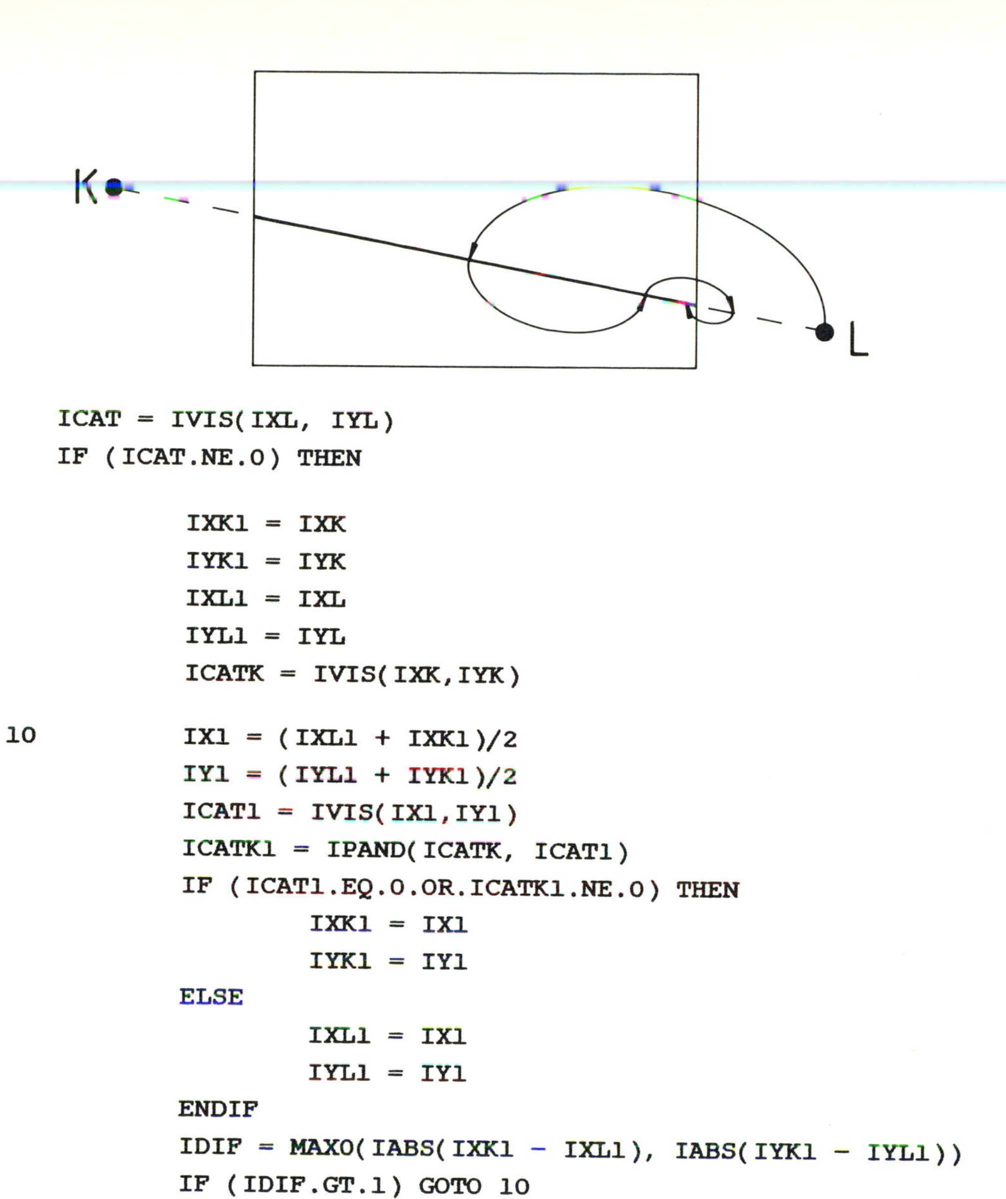


```
      ICAT = IVIS(IXL, IYL)
      IF (ICAT.NE.0) THEN

               IXK1 = IXK
               IYK1 = IYK
               IXL1 = IXL
               IYL1 = IYL
               ICATK = IVIS(IXK,IYK)

10             IX1 = (IXL1 + IXK1)/2
               IY1 = (IYL1 + IYK1)/2
               ICAT1 = IVIS(IX1,IY1)
               ICATK1 = IPAND(ICATK, ICAT1)
               IF (ICAT1.EQ.0.OR.ICATK1.NE.0) THEN
                        IXK1 = IX1
                        IYK1 = IY1
               ELSE
                        IXL1 = IX1
                        IYL1 = IY1
               ENDIF
               IDIF = MAX0(IABS(IXK1 - IXL1), IABS(IYK1 - IYL1))
               IF (IDIF.GT.1) GOTO 10
      ELSE
               IXK1 = IXL
               IYK1 = IYL
      ENDIF
```

Here is the executable code from the function IVIS. The clipping rectangle corners might be passed to this function in a COMMON block.

```
      IVIS = 0
      IF (IX.LT.IXSW) IVIS = IVIS + 1
      IF (IX.GT.IXNE) IVIS = IVIS + 2
      IF (IY.LT.IYSW) IVIS = IVIS + 4
      IF (IY.GT.IYNE) IVIS = IVIS + 8
```

This function, like the rest of the code, would probably be written in low level language as was mentioned above. The same is true for the function IPAND, which can be coded in FORTRAN, but which reduces to a single AND instruction in low level language.

This code finds the point (IXK1, IYK1), which will be the furthest visible point from K. A similar procedure is used to find the furthest visible point from L. If the value of IVIS(IXK1,IYK1) is not zero at the end of the first pass then none of the line segment is visible.

This whole algorithm is quite efficient, as the time it takes only increases logarithmically as the length of the line increases.

Clipping Polygons

The second clipping algorithm that will be described is for clipping a polygon to any convex polygon, though in practice the clipping polygon is usually a rectangle. The algorithm was invented by Sutherland and Hodgeman. It works on any polygon described by a cyclically ordered list of its vertex coordinates. It produces a new ordered vertex list that describes the clipped polygon. The algorithm produces the result shown here.

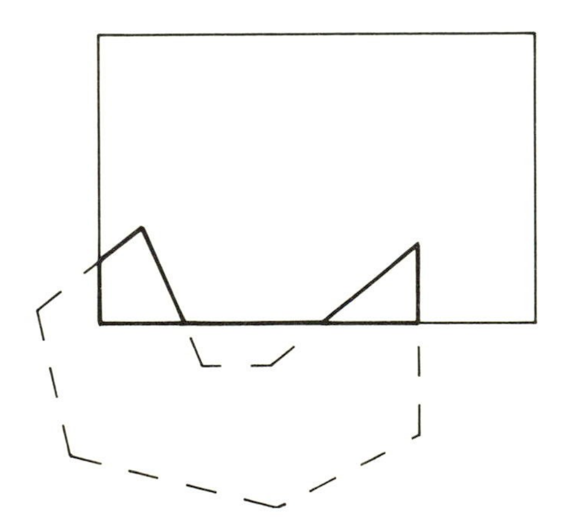

Parts of the polygon off the edge give rise to a degenerate section running along the edge. The algorithim is elegantly simple, and, if the clipping polygon is a rectangle, it can use the line clipping algorithm described above. Otherwise the technique described in Section 3.3 should be used.

Each vertex is compared with each clipping edge in turn. If the vertex is on the visible side of the edge it is added to the output list of vertices. If the vertex is invisible it is discarded, and the line joining it to the previous vertex is clipped to the edge if that previous vertex *was* visible. The clip point is then added to the output list. This last process is omitted for the first vertex, and, at the end of the vertex list, the last vertex is linked back to the first to perform this operation.

The new vertex list is then clipped to the second edge of the clipping polygon, then the third, and so on until all the clipping edges have been used.

As the order of the vertices is preserved throughout the whole process, as soon as a vertex is added to the output list it can be clipped to the next edge, and the next, and so on for all the clipping edges by calling the clipping procedure recursively. This reduces the storage required to hold intermediate vertex lists, but presupposes that the reader is employing a language that allows recursion. Alternatively the vertices may be held in cyclic order in a linked list, when it becomes an easy matter to delete invisible vertices, or to add new ones on clipping edges, while preserving the cyclic order.

References

A. Balfour and D.H. Marwick

Programming in Standard FORTRAN 77

Heinemann Educational Books, 1979 (ISBN 0 453 77486 7)

An excellent introduction to the FORTRAN 77 language, but not a first introduction to computer programming.

I.N. Bronshtein and K.A. Semendyayev

A Guide Book to Mathematics

Verlag Harri Deutsch, 1973 (ISBN 3 87144 095 7)

One of several such guide books. The authors have found it helpful over a number of years. It gives brief summaries of mathematical techniques up to University undergraduate level, and particularly useful tables of information.

I.D. Faux and M.J. Pratt

Computational Geometry for Design and Manufacture

Ellis Horwood, 1979 (ISBN 0 85312 114 1)

This text deals with more complicated geometrical constructions for Computer Aided Design. In particular, it extends the ideas of parametric functions, which we introduce in Chapter Five, to three-dimensional surfaces.

A.M. MacBeath

Elementary Vector Algebra

Oxford University Press, 1964

A good introduction to vector algebra, with particular emphasis on applications in three-dimensional geometry.

W.M. Newman and R.F. Sproull

Principles of Interactive Computer Graphics

McGraw Hill, 1979 (2nd Edition) (ISBN 0 07 046338 7)

This is the standard work on computer graphics. We strongly recommend it to all readers of our book.

G. Stevenson

An Introduction to Matrices, Sets and Groups

Longman, 1974 (4th Impression) (ISBN 0 582 44426 6)

Clearly and concisely achieves coverage of the topics in its title.

G. Stevenson

Mathematical Methods for Science Students

Longman, 1970 (7th Impression) (SBN 582 44424 1)

This is a standard textbook on engineering mathematics. We recommend it because of its excellent introduction to determinants.

G.P. Weeg and G.B. Reed

Introduction to Numerical Analysis

Blaisdell, 1966

Gives details of the errors which result from the use of floating point arithmetic. It also has a section on interpolation.

J.H. Wilkinson and C. Reinsch

A Handbook for Automatic Computation. Vol II Linear Algebra

Springer, 1971

This is a standard work on the numerical solution of systems of linear equations. It is the basis of sections of the NAG library.